ENGINES AFLOAT

From Early Days to D-Day

VOLUME II:
The Gasoline/Diesel Era

Stan Grayson

DEVEREUX BOOKS
Marblehead, Massachusetts

Published by Devereux Books
PO Box 503
Marblehead, MA 01945

Internet address: www.devereuxbooks.com

Library of Congress Cataloging in Publication Data

Grayson, Stan, 1945–
Engines Afloat / Stan Grayson
cm.
Includes index.
Contents: v. 1. The gasoline era — v. 2. The gasoline/diesel era.
ISBN 0-9640070-4-5 (vol. 1). — ISBN 0-9640070-5-3 (vol. 1 : pbk.)
ISBN 0-9640070-6-1 (vol. 2). — ISBN 0-9640070-7-x (vol. 2 : pbk.)
1. Marine engines—United States—History. 2. Internal combustion Engine industry—United States—History.
3. Inboard-outboard engine industry—United States—History. I. Title.
VM731.G684 1999
623.8'7—dc21 99-27121
CIP

On the cover: the one-cylinder marine engine is an Atlas-Imperial with make-and-break ignition that was restored by David Milholland (Photograph by Stan Grayson). The green-painted diesel is a 1940-vintage Atlas-Imperial still operating today aboard a tugboat (Photograph courtesy of Roland Culbertson). The three-cylinder is a Kahlenberg that was lovingly restored by John Egger and photographed by Stan Grayson. The rescue-boat illustration appeared in a World War II era Packard ad.

Book and cover design by Alyssa Morris

Table of Contents

VOLUME TWO

Acknowledgments

While those who assisted the author in this project's Volume I are acknowledged in that book, several who also contributed materially to the chapters in this volume are worth a second vote of thanks. Chief among them were Bill Holleran and Professor Richard Scharchburg at the Kettering/GMI Alumni Foundation, and Paul O'Pecko at the G.W. Blunt White Library at Mystic Seaport.

Those who were especially gracious with their aid on the chapter involving the Navy were Alice S. Creighton and the Special Collections staff of the Nimitz Library at the U.S. Naval Academy, and Duff G. Porter III at the Norfolk Navy Yard. Dr. Val Husley at the Maritime & Seafood Industry Museum in Biloxi, Mississippi, and the State Historical Society of Iowa provided prompt assistance with material that found its way into the chapter entitled "Fishing."

On the West Coast, thanks again to Charlotte McKenna. Others who merit a special vote of thanks include Ed Larson, Dave Peterson, Christopher Seitz, Greg Hicks, Don Koblick, and Steve Phillips. In Oregon, thanks to Anne Witte at the Columbia River Maritime Museum and, in Washington State, to Dave and Dorothy Milholland, Roland Culbertson, to Duff McCandless, Marion Louberbach, Eric Schmitt, John Black, Walt Carruthers and the volunteers at the Northwest Seaport.

Research involving aspects of Atlas included, Dan Webster, Dan Grinstad, Stewart Marshall, John Seymour, and William Worthington at the Smithsonian Institution. The chapter on riverboats and their engines was much enriched by the very special assistance, hospitality, and memories of Bud Daily; by Frank Goulde of the Knox County Historical Society, the Cleveland Public Library, the St. Louis Mercantile Library, and by: George Neargarder, Jack Custer, Paul Striegel, Woodie Rutter, Charlie Stone, Nelson Jones, Jim Poindexter, Bert Shearer, Robert Miller of the Keokuk River Museum, and Dan Owen of the *Inland River Record.* Assistance on Fairbanks-Morse was provided by the Sterling-Rock Falls (Illinois) Historical

Society and by Terrance Ryan and Jim Carr at the Fairbanks-Morse Engine Division of Coltec Industries. Andy Kraushaar and Harold Miller of the State Historical Society of Wisconsin assisted with materials on Busch-Sulzer. Michael Ridderbusch of Special Collections at West Virginia University assisted with background on Ward.

Other individuals who assisted included Ben Fuller, Ernest Felterman, Dick Gibbens, Pierre McGraw, Robert Mayeux, John Kahlenberg, John Egger, Will Cummings, Richards Miller, Don McCloud, one-time PT boatman, Don Mason, and P.T. Boats, Inc. in Memphis, Tennessee. Thanks to Canadians Ray Johnson in Newfoundland and John Budgell in Labrador.

Several researchers assisted me at important times in this volume of *Engines Afloat* including Judith Robinson, Sandra J. Zipperer, Sarah Wales, and Peter Acsay. Finally, again, a special thanks to my long-time friend and colleague, who provided her always expert editorial input, Beverly Rae Kimes.

Introduction

Although there is considerable curiosity about the impact of the internal combustion engine on American working boats and Naval craft, relatively little has been written about either subject. To date, more research has been conducted into traditional small craft powered by sail, oar, and steam than by an internal combustion engine. The earliest days of the Navy's adoption of internal combustion have likewise received comparatively little attention. This volume focuses on both areas. Both represent fascinating areas of study, and both are challenging to research.

One senses that events moved at a rapid pace throughout the early days of the marine engine's introduction to work boats and the Navy, and that thorough documenting of machines that were soon taken for granted was left undone. The result has been an absence of information about some fascinating technology and those responsible for it. Even more rare than first-rate research sources are photographs of those first engines as installed in the first generation of gasoline-powered towboats, fishing schooners, launches, and the myriad other little vessels that so revolutionized the way men worked on the water.

Volume II of *Engines Afloat* focuses primarily on the impact of the internal combustion engine on fishing and commercial vessels, and on the adoption by the Navy of such engines. Initially, it was the gasoline engine that either augmented oar- and sail-driven working craft or replaced steam engines. The result was a far-reaching revolution in the way that boats were designed, operated, and manned. The subsequent shift from gasoline to diesel was driven by the twin needs for lower operating costs and improved safety. At first, it was the biggest gasoline engines that were replaced but gradually the shift came in the smaller horsepowers as well. It was a sometimes bumpy transition made possible only by stubborn, often brilliant men who pushed the limits of technology with unending perseverance.

Long before the marine diesel engine was made practical, however, the gasoline engine had already forever changed the nature of life for those who got their living on the water. Fishermen, whether on the coasts, Great Lakes, or rivers, were quick to recognize the backsaving, lifesaving, and economic potential of the machines. For the men who worked on the inland rivers, the gasoline engine spelled an end to the domination of steam commerce with its many regulations and high expenses. They, along with those who operated tugboats all along the coasts, quickly began adopting the internal combustion engine. Although it did not alter the fundamental nature of most waterborne occupations, the gasoline engine impacted in a generally positive way almost all the details of how that work was accomplished. Such engines were a blessing and people stuck with them until first the semi-diesel and then the true diesel became reliable alternatives. Then, most everybody made the switch.

It was, in retrospect, a fascinating time. On the rivers, in particular, the adapting of big, stationary engines to drive a paddle wheel was touched by the genius of men who possessed little education but who simply knew how to do things. In the Navy, the idea of replacing steam-powered launches with gasoline versions took some years to find acceptance. Along the way, a variety of fascinating engines were procured and then produced in the Navy's own shops. But more than small craft were involved. One can only wonder today at the spirit of those who designed the first gasoline submarines, and of those who actually dared to become shipmates in them. Four decades later, however, gasoline engines like Packard's V-12 and diesels from Fairbanks-Morse and General Motors would become, in the service, legends in their own time.

Machines like the World War II-vintage Fairbanks and GM's 6-71 and other models were not even dreamed of in the early days of diesel development. Then, there was the struggle to simply perfect a reasonably compact, reliable engine whose fuel could be injected by some means other than high-pressure air. There are men still living who remember these machines from the great "iron age" of the diesel engine, and the thoughts and memories of several are included in this volume. The diesels they recall so nostalgically were slow-turning designs which, were it not for the ravages of salt water within their cooling systems, might have lasted seemingly forever. Some do, in fact, survive both in museums and even aboard boats. These gentle giants had their beginnings as ungainly machines that evolved into the stalwarts known to an entire generation of enginemen.

The demise of these big cast-iron diesels and their almost universal replacement after World War II was among the most dramatic of mechanical transitions. It saw the end of engine-building companies whose names were household words on waterfronts throughout the country and it witnessed the end of all but a few American diesel manufacturers. Afloat, the new-style diesels had been the engines that drove what General Eisenhower always called "the little boats that won the war." They were the diesels that powered craft as diverse as the Navy's landing craft and fleet submarines. On land, the new generation of lighter, faster-turning diesels made practical the first modern diesel locomotives. After the war, surplus GM 71-series diesels replaced almost all the old-generation engines on boats that serviced Gulf Coast oil rigs and then, all the others besides.

As in Volume I, a comment is in order regarding the time span noted in the subtitle — "from early days to D-Day." The primary focus is on the years 1884 - 1944. Although Volume I discusses events before 1884, they are included there as important background. The real story of the marine engine industry in this country can be most reliably defined as beginning in 1884 with the application of a Regan engine to a launch in San Francisco. Forward-thinking commercial boatmen were quick to perceive the promise of internal combustion, and they formed a key early market for the first engine makers.

Today, some may find surprising the early date at which a few companies, with the commercial operator foremost in mind, began development of engines that would operate on comparatively inexpensive fuel oil rather than gasoline. And, if today, the diesel engine is taken for granted, it is well to pause and reflect on the labor, brain power, money, and risk required to perfect it. This human and financial capital was invested over a period that is measured not simply in years but in decades. In fact, the quest begun by Diesel himself spurred a movement that remains in-process still. But it was the period from World War I to World War II that resulted in what were, perhaps, the most dramatic chapters in the diesel story. They were chapters written by men as diverse as the little-known A. C. Estep in Seattle, and the famous Scots-born, self-made millionaire in Ohio, Alexander Winton. There was the patient and innovative Clessie Cummins, and the great engineering and scientific genius of his era, Charles Kettering. Their stories and others are told in the following pages.

Stan Grayson
Marblehead, Massachusetts
June 1999

Gasoline/Diesel Marine Engine Development: Periods and Events

Pioneer Period (1884 - 1898)

1884	• First Regan engine; soon adopted by San Francisco launch operators
1890	• Sternwheeler w/gasoline Regan engine built at Scotia, CA • Charter gasoline engine production begins in Sterling, IL
1892	• Diesel patent
1893	• Charter-Fairbanks gasoline engine built at Beloit, WI
1895	• First engine built under Fairbanks-Morse nameplate
1897	• Adolphus Busch acquires U.S. patent rights for Diesel's engine
1898	• First Kahlenberg, Two Rivers, Wisconsin • Globe gasoline engine fitted to Gloucester schooner *Helen Miller*

First Production Period (1899 - 1912)

1900	• Launch of Navy's first submarine, *U.S.S. Holland* powered by Otto gasoline engine • Fairbanks-Morse introduces Type B marine engine
1903	• St. Marys horizontal stationary engine sees adaptation aboard river "gas boats"
1905	• Gray Motor Co. founded in Detroit
1906/'07	• First four-cycle Sterling produced, Buffalo, NY. "Navy" models follow
1909	• Electric Boat builds E and F Class submarines w/Vickers diesel engine
1912	• First U.S. diesel-powered ship built at Nelseco w/M.A.N. diesel • Winton Gas Engine and Manuf. Co. established, Cleveland
1913	• Winton builds first "American-designed" diesel • First Nelseco-designed diesel powers cannery tender *Warrior*
1915	• First Atlas-Imperial diesel • Winton builds first V-12 diesel

Maturation Period (1919 - 1930)

1919	• Atlas-Imperial introduces common rail fuel injection
1920	• Enterprise builds its first diesel in San Francisco
1921	• Washington-Estep production begins in Seattle
1923	• Towboat *Harvey* debuts with direct reversing diesel

1924	• Fairbanks-Morse debuts new marine diesel, Model 32
1925	• Winton is premier supplier of diesels for yachts.
1926	• Buda Company acquires M.A.N. patent for high-speed diesel.
1928	• GM begins diesel research/development • Cummins introduces fully enclosed Model "U"
1929	• Cooper and Bessemer merge to become Cooper-Bessemer

Prewar Modern Era (1930 - 1939)

1930	• First GM unit injector • GM acquires Winton • GM acquires Electro-Motive
1931	• Launch of "World's largest river towboat" *Herbert Hoover* w/screw propellers and McIntosh & Seymour diesels
1933	• GM displays unit-injected V-12 diesel at World's Fair • U.S. Navy announces it will "dieselize" all boats
1934	• Navy acquires license to build Buda diesels
1936	• Cummins announces a six-model line of high-speed diesels
1937	• GM forms Detroit Diesel Engine Division
1938	• Gray introduces marinized GM 6-71 as Gray Marine diesel

World War II Era (1939 - 1944)

1939	• Packard receives contract for 81 4M-2500 V-12 PT-boat engines
1940	• Detroit Diesel begins production of diesels for tanks and later marine applications — by war's end, over 182,000 diesels produced • Navy becomes sole consumer for Fairbanks-Morse production of opposed-piston models
1941	• GM's Electro-Motive steps up production of 567 diesel for LST and other uses
1942	• GM Model 84 "pancake" diesel approved for production
1944	• Gray Marine (GM)-powered landing boats assault Normandy beaches on D-Day

Note: Volume I includes a timeline that focuses more specifically on dates important to gasoline engine development, its technology, and its impact on recreational boating and racing.

Atlas-Imperial Diesel Engine Co. plant, Mattoon, Illinois, circa 1940. Atlas opened a Midwest facility in the mid-'30s and built a wide range of models there. In addition, during World War II, munitions were manufactured. (Courtesy, Dan Webster)

Chapter 1

Diesels

"My first engine was constructed in 1893 but never succeeded to run, not even one revolution, but at the first injection of fuel there occurred a terrible explosion, nearly killing me."

—Rudolf Diesel, 1912

"In the diesel field we note the trend toward higher speeds and greater horsepower per pound of engine weight. Yearly the diesel assumes more of the characteristics of the gasoline engine."

—*Motor Boating*, January 1938

"When I came to Juneau in '37, there were still many slow-speed gasoline engines. The first Alaska fishboat with a diesel that I saw was a salmon seiner whose owner couldn't afford a new gas engine and put in a used Caterpillar. It was a big success. But my personal favorite was the Atlas-Imperial."

— H. R. "Slim" Blood in an interview with the author, April 1996

The wooden tugboat, *Quail*, lay in her slip in D Dock at Anacortes, Washington, her decks wet with the intermittent rain that had been falling for several days. All around were modern boats, stubby aluminum seine skiffs — glinting dully in the occasional breaks of pale sunlight — and big, steel towboats, but the *Quail* seemed right at home. Over a half-century old, this was a boat that still belonged in a working harbor. The *Quail* was built in 1940, at a place then known as Kogging, Alaska, on the Kivchak River. It was then, as now, a remote area of big tides, and Kogging is no longer even a dot on the map. But, for many years before World War II, this region, near the northeast corner of Bristol Bay, was busy with salmon fishermen.

During the off-season, the multiskilled men who worked for the Alaska Packers Association were set to work building and maintaining the many boats that served the company's canneries. These vessels were built (of lumber shipped up on barges from Seattle) right by the river, which was reached by an extensive marine railway. Although vessels like the *Quail* still turn up on occasion in the Pacific Northwest, they exist today mainly in photographs. One sees them in the old boating and shipping magazines or else in pictures hanging on the walls of well-established towing companies. They are sturdy looking, evocative vessels that once did the work of a whole region, but most of them are gone.

So, it is a pretty rare thing to step aboard the *Quail,* a boat with a black hull, white cabin sides, and neatly painted blue trim. The vessel is 73 feet long, has a 17.5-foot beam, and draws eight feet. She is built of thick fir planks over sawn fir frames, the whole fastened together with big, galvanized iron ship spikes and bolts. Perhaps the most unusual thing about the boat is her engine. Down below, at the very heart of the tug, stands the very same power plant with which she was launched. How this engine survived for so long is something of a miracle. It is a machine from the "Iron Age" of marine engines. The engine is painted a pleasing sort of medium green and it smells slightly of oil. It is a six-cylinder Atlas-Imperial, an engine that was once a fixture all up and down this vast, tree-fringed coast. The cylinders of the big Atlas, with its 10-inch stroke and 13-inch bore, generate a mighty 200 horsepower.

Six-cylinder Atlas-Imperial diesel, 1940. This 200-hp Atlas is the classic model with pushrods and rocker arms fully exposed. It powers the 73-foot tugboat Quail. *(Courtesy, Roland Culbertson)*

"Most of these old diesels were yanked out," said Roland Culbertson, who has owned the *Quail* since 1960. "But with the *Quail,* they hauled her out in 1948 and replaced the saltwater cooling system with a keel cooler. So saltwater never had the chance to dry out in the water jackets and wreck the engine."

The Atlas stands as tall as a man and its six individual cylinders are bolted to a crankcase that extends for some 16 feet in the center of the engine room. On the aft end is a big flywheel that encloses a sailing clutch so that the engine, which is direct-reversing, can also be moved in and out of gear. By 1940, there was a well-established trend to enclose the working parts of big diesels. But this particular Atlas is old school. Everything is out in the open, an amazing conglomeration of valves, rods, levers, and pipes. Once you know what each does, however, the mechanical aptitude that the Atlas represents is less bewildering than remarkable. Poking up from the crankcase along each cylinder are four push rods. Like everything else about the engine, the rods look sturdy, thick as a man's thumb. Two operate the valves; the third operates an air valve, and the fourth works the fuel injector.

Roland has decided that, as a demonstration, he will start the Atlas. Today, he skips the one aspect of the starting drill that might be called a nuisance. An oiling can sits near the engine, and normally it would be used to apply a dollop of oil to each of 160 spots. These, in turn, would require an additional oiling every two to four hours. "That's one thing you didn't have to do on an enclosed engine," Roland noted.

He oiled just a few spots, mostly on the big push rods. A control wheel and quadrant are mounted on the engine's starboard side. The quadrant is engraved with the words, "Reverse," "Stop," and "Forward." Roland turns the wheel to the Forward position. It is an action that sets off a remarkable chain of events. Plungers and rods begin to move in a predestined sequence, sending air into the engine's cylinders at some 225 p.s.i. The air is followed quickly by fuel that is pumped through a line ("rail") to the fuel injectors at pressure that can reach 4,000 pounds p.s i. or, if need be, up to 7,000 pounds. It is this mechanical fuel injection system, any diesel man will tell you, that made Atlas great. Within about five seconds, the Atlas is running. It does so with a steady, reassuring beat and it is surprisingly quiet.

Two gauges are mounted on the engine; an air pressure gauge and an oil pressure gauge. There is no tachometer. "Don't need one," said Roland. "You just do this." He reaches up and lightly grasps one of the push rods. "You just use the second hand of your watch to time things and, pretty soon, you can tell just from experience. Normally, you might run at about 300 rpm. But it will turn up to 350 or 400 rpm if you need it."

The boat's early career as a cannery tender was cut short by World War II, and so the *Quail* spent her youth hauling for the Army and then became part of the fleet that supported the building of the DEW (Defense Early Warning) line. "She burns eight to nine gallons per hour at rated horsepower," said Roland, "and a Cat or Jimmy uses double the fuel to turn the same size propeller." This is not an idle boast. Although he now keeps the boat as a kind of hobby and to stay active in retirement, Roland worked the *Quail* hard, so he had ample chances to compare the boat's performances. After a while, Roland shuts down the engine. It stops as quietly and with as little fuss as it started.

In the now still engine room of the *Quail,* the old Atlas seems somehow amazing. We are standing beside a piece of machinery whose essential design features were worked out by 1919, yet the engine is still ready, in a few short minutes, for a day's work. It is exactly as described in the Atlas catalog: "The engine is massive, simple in construction, all parts easily accessible, all parts as far as possible interchangeable. Engine can be operated by any ordinary intelligent individual. All parts where practical are hardened and ground to precision gauge."

Over the years, Roland has rebuilt the fuel injectors and ground the valves, easy to do because the latter are mounted in easily accessed "cages," and that's about it. With its freshwater cooling system, the *Quail*'s engine gives every indication of being able to last for a hundred years, perhaps much longer. And this isn't even the oldest Atlas in the neighborhood. There are others still running that are a decade older.

The Atlas-Imperial Engine Company of San Francisco was the result of a merger between two of the city's engine builders, the Atlas Gas Engine Company and the Imperial Gas Engine Company. Atlas had commenced manufacture in 1904 and it is reported that Imperial was begun at about the same time. Atlas was started by three men who had previously been working at one of San Francisco's pioneer companies, the Hercules Engine Company, which was in business in the early 1890s and which would eventually supply engines for applications as diverse as boats, buses, an Alaskan sleigh and even to drive vacuum cleaner systems in city hotels. In 1895, three key men left Hercules. They were factory superintendent Jack Lorimer, design engineer A. Owens Warenskjold, and draftsman Peter Bahr. This was the trio that formed Atlas.

Factory of the Atlas Gas Engine Company, Oakland, California, 1917. The name reflects Atlas's beginnings as a maker of gasoline engines. Its first diesel had been completed in 1916, before this photograph was taken, the year in which the company joined forces with the Imperial Gas Engine Company of San Francisco. Their goal: convert production from heavy-duty gasoline engines to diesels. (Courtesy, Oakland Public Library)

Exactly what happened to cause these individuals to leave Hercules is now unknown. But Lorimer and Warenskjold, at least, were likely present when Hercules was founded in 1889 or 1890. When they started their own company, they reverted to Greek mythology for its name. The name they selected was perhaps the only one that could rival Hercules for conveying an image of great strength — Atlas.

The Atlas Gas Engine Company, makers of stationary, portable, and hoisting engines, had just established itself when the San Francisco Earthquake struck in 1906 and destroyed the entire fledgling enterprise. In common with most within the devastated city, the response of Atlas's founders was to seek additional capital and to rebuild immediately. The money came from Marshall Harris of the American Dredging Company, who assumed the duties of company president. A new sales office was opened on Market Street and a new factory was established in Oakland at 11th Avenue and East 9th Street.

"One of the largest factories located on Oakland harbor" is how the plant was described in the newspaper in January 1909, "occupying a site comprising fifteen acres." By then, the company's annual production was generating close to a million dollars in revenue and a 75-man work force was building marine and stationary engines of four to 500 horsepower that were already being shipped as far away as Australia and, of course, to the Pacific Northwest and Alaska. "Agencies," the company said proudly, "are maintained in almost every point of consequence in the entire world with the exception, perhaps, of Europe…."

Interior, Atlas Gas Engine Company, 1917. Gasoline engines were still the mainstay of Atlas-Imperial when this photograph of the shop floor was made. (Courtesy, Oakland Public Library)

Three-cylinder Atlas gasoline engine, circa 1910. Atlas had begun, and prospered, as a maker of heavy-duty marine engines. Models offered included one-, two-, three-, and four-cylinder versions ranging in power from 4½ to 200 hp.

ıtalog.

ʏlinder types.

Beginning in its earliest days, Atlas seems to have been a particularly well organized and managed company. Said an early catalog: "Each department working as a separate unit, yet merging into one harmonious whole; the finished product a direct result of successfully applied modern mechanical skill, modern factory methods and modern machinery usage." As the company grew, it adopted an open-minded approach to offering workers the opportunity for change and growth. According to Glenn Parker, who joined Atlas around 1920 and eventually became responsible for the installation of engines in the field, employees could train for other jobs within the factory on company time. When it was determined that sufficient skill had been developed, a job change could be made. A man injured on the job and no longer able to perform a given task, had the opportunity for retraining for a new position. These were important innovations in an industry not unknown for union-organized strikes.

The engines were all heavy-duty machines — "the Gas Engine That's Surer Than Steam" — that found their way into a variety of workboats, fishing boats, and some pleasure craft and earned a reputation for robustness and reliability. They were make-and-break ignition, T-head motors, with cams, valves, and other components arranged for easy access. Typical of the service rendered by these big Atlas gasoline engines was the 150-horsepower model installed in the *Empress,* owned by the Lauritzen Transportation Company of Sacramento. By the time the engine had driven the boat for some 35,000 miles, a sum of nine dollars had been spent on repairs.

Meanwhile, over in San Francisco, on Tehama and Clementina Streets, the Imperial Gas Engine Company was building engines that ranged from single-cylinder models to 200-horsepower, six-cylinder versions. They were overhead-valve engines, neatly constructed, and they were being sold to a market that seems to have been the very same one that Atlas served. "The Reliability and Power of the Imperial has Made it the 'Marine Draft Horse' of San Francisco Bay" said an ad. One ad pictured the scow *Chas. W.* carrying 165 tons of hay across the Bay. The boat had a pair of 30-horsepower Imperials. Atlas and Imperial each had its own network of sales agents and both companies would seem to have been doing reasonably well in 1916 when a decision was made to join forces, forming the Atlas-Imperial Engine Company.

Four-cylinder Imperial gasoline engine, circa 1907-'08. The West Coast market for heavy-duty marine engines was big enough to support Atlas and Imperial as well as Union, Standard, and Hicks. This handsome Imperial displays a marvelously conceived make-and-break ignition system and comprehensive oiling system. (Courtesy, San Francisco Maritime Museum National Historic Park)

Although the manufacture of gasoline engines continued under the Atlas-Imperial name, the real reason for the merger was a desire to build and sell diesel engines. This could well be interpreted as a merger dictated by a "desire to survive" because those who remembered that period recalled it as a time when "something had to be done." Atlas had begun development of a diesel engine in 1914, a time when gasoline prices had risen and big gasoline engines were becoming comparatively expensive to operate. Semi-diesel engines that ran on crude oil were already beginning to make inroads into the gasoline engine market. What's more, papers describing the progress of the diesel in European workboats, especially tugs, had begun to appear with some frequency.

In 1914, an article in *International Marine Engineering* suggested that a towboat or tug would derive substantial benefits by converting from steam to diesel. Using then-current prices of 20 cents per gallon for gasoline and five cents for diesel, the author of "Diesel-Engined Tugs," J. Rendell Wilson, calculated a yearly fuel cost of $12,672 for a gasoline tug working 330 days a year versus $1,775 for a diesel. This savings would soon pay back the greater purchase cost of a diesel engine, which he gave as $12,200 for a 120-horsepower,

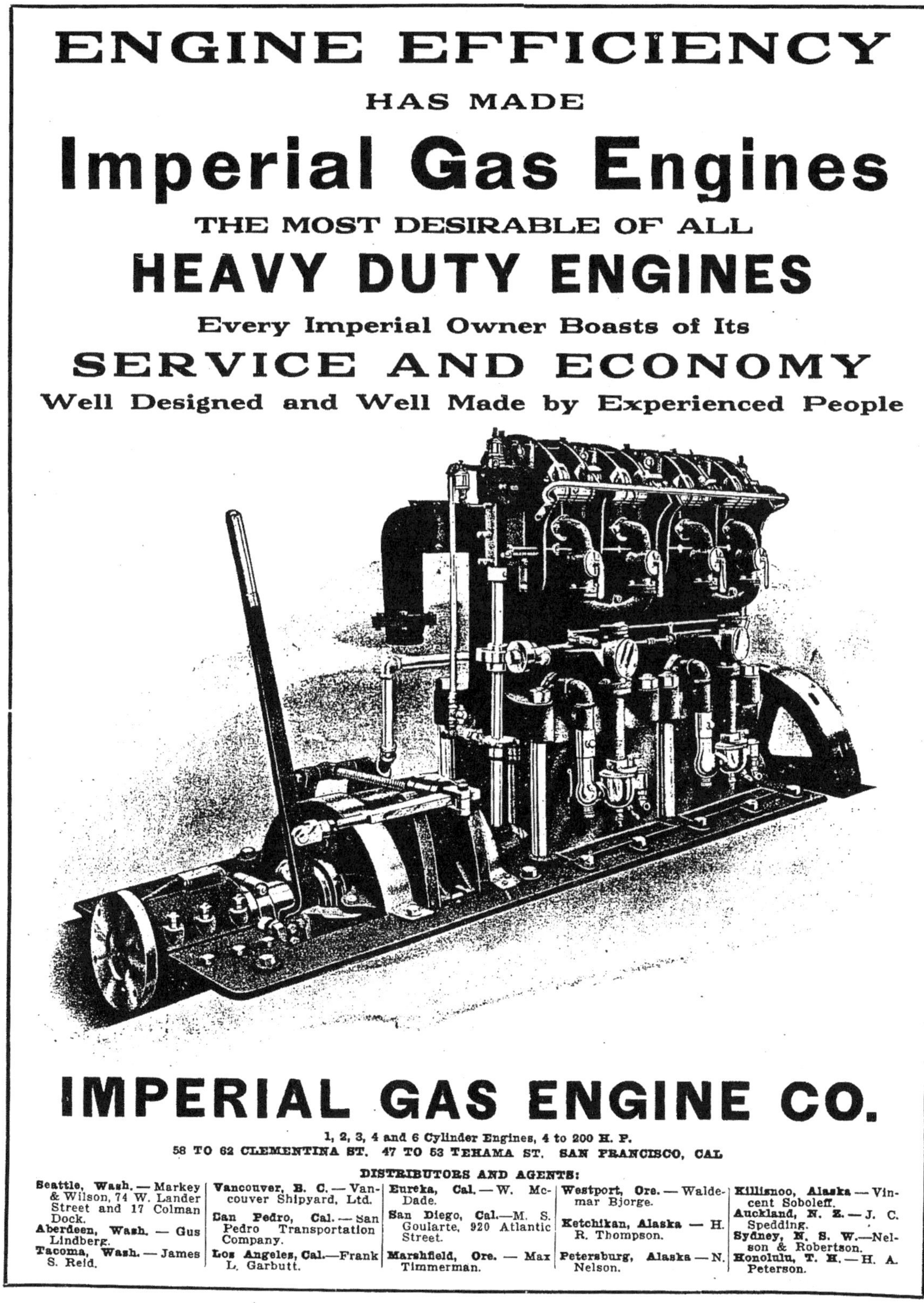

Imperial Gas Engine Company ad, circa 1912.

four-cylinder diesel engine and its associated air compressor and tanks. Such boats were then operating in Europe but Wilson noted that before they could become common in the United States, "the manufacture of these motors will have to become more general in America, and better arrangements for distribution of heavy oil fuel will have to be made by the oil companies." Atlas was among the companies that wanted to lead in making the availability of a diesel "more general."

The first Atlas diesel was an experimental, four-cylinder machine with a nine-inch bore and 12-inch stroke. It was rated at 110 horsepower but was found to produce 150 when run at 300 rpm, a higher speed than its normal rpm. Fuel was injected by air pressure as then was typical. Consumption of .446 gallons per horsepower hour was considered gratifying but not phenomenal. Still, it was certainly a successful first try, and Atlas-Imperial engineers began working on a big, six-cylinder engine that, with an 11½-inch bore and 14-inch stroke, was expected to develop 250 horsepower. This engine was actually designed for a customer, a mining operation that promptly canceled the order with the outbreak of World War I. Instead, in 1916, the engine was equipped with a flywheel and one-way clutch on each end and installed on the Seattle ferryboat *Vashon Island.* Thus began a long career for "Atlas Number 1." The ferry's name and route were changed twice, first to *Mercer* and later, in 1939, to *Islander.* At that time, the boat was operating between San Pedro and Terminal Island, California. This boat, and its Atlas diesel, were still in service after World War II.

For the first three years of its service life, the big Atlas diesel operated with its air-injection fuel system. But in 1919, Atlas came out with a development that changed the basic nature of its engine, and the entire way in which the marketplace perceived diesels. The need to use high pressure air to deliver and atomize the diesel fuel was a stumbling point for engine builders, both European and American, and for their customers. The system required a two- or three-stage air compressor, heavy air tanks, and piping and fittings stout enough to withstand 1,200 pounds of air pressure. "High pressure air did a beautiful job of fuel atomization," said "Slim" Blood, who operated such engines as a young man, "but keeping the air lines tight was a problem that was never really overcome. They didn't have the materials."

In its catalogs, Atlas would refer to such compressors as "costly, dangerous and troublesome." They also robbed power and fuel efficiency. As the demand for easier-to-operate diesels began to grow, it was recognized that air injection could not be tolerated. Atlas specialist Dan Grinstad noted that on air-injected diesels, "Pipes would burn the paint off and glow in the dark. You couldn't leave the engine room, and running such an engine required a lot of finesse by comparison."

The comparison was to an engine that did not need high pressure air for its fuel system and, in 1919, Atlas-Imperial, introduced one. "As a result of its desire to get away from this [air injection] feature," reported *Diesel Progress* in May 1948, "the company developed by its own efforts a fuel injection system of its own (although Vickers in England had developed and used such a system) according to 'Common Rail' injection system. So successful has this system been that it is still in use with a few modifications in details, after some 27 or 28 years."

According to Atlas engineer Roy Anderson, the company actually built only three or four air-injected models and these were all eventually converted to the common rail injection system. The company did not publicize details of how its common rail fuel injection system was actually developed. But, although Atlas pointed with pride to its diesel as being "all American," it appears this was not literally so. A. C. Estep, who worked for several years in Seattle for the Atlas-Imperial agent there before developing his own diesel engine, reported in an interview that, after World War I, an engineer named George Adkin "who had worked in England now joined their staff.... Atlas followed along with the English idea on the principle of a high pressure oil spray into each cylinder."

The "common rail" simply referred to the fuel line that supplied each of the injectors or "spray valves" with fuel pumped from the tank through a strainer that would hold back any particles larger than the tiny holes — 0.008 to 0.016 inches in diameter — in the spray valves' tips. Fuel pressure was developed and maintained by mechanical pumps to a level as high as 3,800 pounds at normal loads. The company explained it quite lucidly:

> "The pressure is maintained by plunger pumps driven by cranks mounted on the cam shaft. These pumps deliver fuel oil into the high pressure system leading to [a] fuel spray nozzle in the center of the cylinder heads. A branch pipe connects each spray nozzle to the high pressure oil system. An additional branch leads to a pressure regulator and relief valve, which regulates the pressure, and by-passes all oil not required to maintain constant pressure.... The spray valve is mechanically opened by a cam and properly TIMED with the cycles of operation to gradually inject the fuel at the proper time for ignition. NO HIGH PRESSURE AIR is used to force the fuel into the cylinder."

Roy Anderson called Atlas "the first company west of the Mississippi River to develop a commercially successful solid injection fuel system." In fact, Atlas may have been the first U.S. diesel engine maker anywhere to make a mechanical injection system *standard* equipment. Vitally important to the engine's great success, and something that long bedeviled those working to develop the lightweight, high-speed diesel, was the fact that it could run on the only available fuel oil, which was high in asphalt content. As a matter of fact, that oil caused problems for semi-diesels, too.

Almost immediately, Atlas began emphasizing the fuel efficiency and reduced operating costs that had always been the goal of its diesel development program. "By our present injection device we are able to obtain the highest economy in fuel," said the company in its 1922 catalog. "Our mechanical injection Diesel engines will deliver the same amount of power with a cost of fuel and lubricating oil of approximately $10.00 which a gasoline engine will deliver for $100.00, in other words the cost of operating our Diesel engines is about 10% of the cost of doing the same work with gasoline. Our engines will deliver 19 H.P. per hour from one U.S. gallon of crude oil and the cost of fuel and lubricating oil is about 25c per hour for 100 H.P. Can you afford to use a distillate or gasoline engine using $2.50 per hour to do the same work that one of our Diesel engines will do for 25c?"

The market, as Atlas-Imperial saw it, was chiefly going to be for marine diesels up to 500 horsepower. It initially developed three cylinder configurations. They included a four-cylinder, 9 inch x 12 inch of 110 horsepower; a four-cylinder, 10 1/2 inch x 13 inch of 125 horsepower; and a six-cylinder, 12½ inch x 16 inch, of 350 horsepower.

Everything that Atlas said regarding the efficiency of its engine was true. Soon, other manufacturers would introduce mechanical fuel injection systems of their own and, for well over a decade, the heavy, slow-turning

Atlas diesels. These two engines, a three- and a four-cylinder, were built at the Atlas factory in Mattoon, Illinois. (Courtesy of Dan Webster)

diesel would rule the marine engine market. Yet, the engines contained the seeds of what would prove to be their ultimate undoing. Their low rpm meant there was no chance to significantly develop higher horsepower by increasing engine speed while, at the same time, reducing weight. The fuel injection system lay at the core of the problem. Years later, in 1928, George Hallett, who was then head of the Power Plant Section at General Motors' Research Laboratories, reported on tests of an Atlas diesel built under license by the Bessemer Gas Engine Company. "One problem in this engine," he reported, "was that the moving parts of the injection system were so heavy that at high speeds their inertia kept them from following the driving cam, and fuel metering accuracy was no more." Developing a higher rpm, lighter-weight successor would prove to be a problem that would require both significant capital and the world's best engineering talent to solve.

By 1914, when Atlas-Imperial began work on the diesel engine that would represent its future, Rudolf Diesel had been dead for less than a year. Born in Paris in March 1858, he graduated from Munich's Technical University in 1880, having achieved the highest marks since the 12-year-old school had opened its doors. Although Diesel and his parents had fled Paris for London in 1870 during the Franco-Prussian War, he returned to France after graduation and began work as a refrigeration engineer. That same year, he was granted a patent for his process of making clear ice suitable for use in restaurants.

Artistically and musically gifted, brilliant and innovative, Rudolf Diesel was also a close observer of society and of his fellow man. He eventually published a little-known book, *Solidarismus,* setting forth his socialistic ideal of a Germany that would offer workers a national social security plan. This book was of more importance to Rudolf Diesel than one might guess. He later told one of his sons that "It is all very well that I invented the Diesel engine, but it is much more important that I have solved the social question."

Rudolf Diesel (1858-1912). Widely recognized for his brilliance, Diesel was also controversial, and his career was dogged by financial and legal setbacks. The patent on which his engine was based was issued in 1892.

Diesel hoped to reduce the burden of labor that he saw wearing down poorly paid factory workers, and he nourished for years a dream of replacing the steam engine with a compact machine that would be economical to run and affordable by a small factory owner. The seeds for his idea of replacing the steam engine with something more efficient dated back to his days in Munich and one of his professors, Carl Linde, inventor of the ice-making machine, and the man who would give Diesel his first job. Linde had pointedly criticized steam power as wasteful of energy, returning only about 10 percent of the heat contained in its coal fuel to energy.

Diesel's creation of a more efficient engine was destined to be a long and arduous process and it took years for him to realize his fuel of choice, initially, was incorrect. Paradoxically, the man who would invent the compression ignition engine with its great heat requirements thought that ammonia gas, the same used in Carl Linde's ice-making machines, could fuel his engine. Diesel spent five years and all of his savings on the project, which always seemed on the brink of some kind of breakthrough that never happened. And always, he suffered bouts of nausea, headaches, burning eyes, and a sore throat caused by the ammonia that attacked all metal but copper and seeped out through invisible holes in the engine's feedlines and parts. Once, an ammonia container blew up. Finally, he switched his attention to coal gas and then to liquid fuel. "How the basic idea came about, replacing ammonia with a real gas, namely highly compressed air at a high temperature and gradually introducing finely divided fuel into this... then letting it expand upon combustion of the individual fuel particles in such a fashion that as much heat as possible becomes energy, I don't know," Diesel wrote later.

Rudolf Diesel, who was a graduate mechanical engineer, began referring to himself as a "thermal engineer" around 1890. He had immersed himself in the isothermic theories of the pioneer Sadi Carnot and had closely studied Daimler's Otto-cycle engine. But Diesel wanted to use a fuel less volatile than gasoline and cause its ignition by compressing it and thus heating it far more than the Daimler's 4:1 compression ratio could ever permit. He had the idea well enough developed to receive a patent on his "rational heat engine" in 1892.

The patent described a "working principle for internal combustion engines, characterized by pure air or a mixture of other indifferent gas and pure air being compressed by the working piston in a cylinder to such a degree that the temperature produced is far higher than the ignition temperature of the fuel to be employed, whereupon fuel is gradually admitted after dead center, so that combustion, owing to the back travel of the piston and the consequent expansion of the compressed air (or gas) takes place without a noteworthy increase of pressure and temperature, the gas in the working cylinder then being further expanded upon completion of the fuel feed."

In January 1893 at the urging of his wife Martha, Diesel published his theories in a booklet entitled, *Theory and Construction of a Rational Heat Engine to Replace the Steam Engines and the Currently Known Combustion Engines.* This little publication, Diesel's first patent, and a subsequent one initiated a conflict with other inventors and engineers who questioned the real meaning, the originality of the claims, and ultimately whether the diesel engine even met the description of it in Diesel's own patent. Although the German Patent Office upheld Diesel, he actually settled with one challenger, Emil Capitaine, and debate would continue for several decades to swirl around his claims and whether his engine as built actually represented the patent. His first publication generated comments that Diesel recalled later were "generally very unfavorable, or, strictly speaking, crushing."

By now, Diesel knew that he could never afford to continue his research and development without meaningful financial investment. In April 1893 he went straight to the 300-room villa of the biggest industrial man in Germany. "The whole of my engine," Diesel told Fritz Krupp, "must be made of steel." He showed Krupp his drawings. Eventually both the steel baron and the Machinenfabrik Augsburg, which had previously turned Diesel down, agreed to assist Diesel with financial and technical aid. In February 1893, a contract was drawn up between Diesel and Augsburg, which was granted manufacturing rights and sales rights for southern Germany. Then, work commenced on making the cast steel piston, the iron and steel cylinder, and other parts for a test engine. In July 1893, the engine was put together in a walled-off area in the plant. The machine stood 10 feet high and had a 5.91-inch bore and 15.75-inch stroke. Even that stroke was not enough to deliver on the desired 30:1 compression ratio. Instead, because of insufficient valve and piston ring sealing, an 18:1 ratio was achieved. Diesel and a long-time engineering friend, Lucian Vogel, went ahead anyway. In August, the engine's flywheel was hooked up to a belt driven off the factory's shafting and when Diesel operated the fuel pump to send a charge of benzene into the engine, a single, very loud explosion blew apart the cylinder head and its compression indicator. The shrapnel embedded itself in the walls and ceiling around the engine.

Still, Diesel had caused an explosion within the engine's cylinder without an ignition source. "I knew then just what I wanted," Diesel said later; "it was possible to compress pure air to such a high degree that the fuel injected into it ignited and burned."

They rebuilt the engine. It coughed and barked and filled the *Machinenfabrik* with black smoke. Valves and pistons burned. Diesel persevered. After over a month had passed, he noted that it would be best to rebuild the engine with a new piston and combustion chamber, a water-jacketed cylinder, different valve layout, and with compressed air used to inject the fuel, which this time was petroleum. These changes resulted in more or less regular ignition but the engine was kept running by the factory's power-takeoff belts. It was only after ongoing adjustments that, on February 17, 1894, Diesel got the rebuilt engine running for a few minutes at a time, under its own power, at about 88 rpm. It was, perhaps, the end of the beginning stage of Diesel's development. The two Augsburg mechanics who'd been assigned to work with him, Hans Linder and Friedrich Schmucker, were somewhat awestruck to witness the most successful demonstration yet of an entirely new invention. As for Diesel, he called himself "a lucky mushroom," reported on his results to Krupp, and went back to work.

There was plenty of work to do, too. Injection of the fuel was a major obstacle, as was ignition. At one point, Robert Bosch was summoned and a magneto ignition system was added. It was not until an engine-driven air pump was added to another design evolution in March 1895, that the engine began to run reliably and, at Martha Diesel's suggestion, was named the "Diesel Motor." Then, a wholly new model, the third one, was built and, by early 1897, it had passed endurance tests of 17 days in length. Running on a grade of kerosene used for lamps, the engine was demonstrated to Krupp's chief engineer and to representatives of the Swiss steam-engine manufacturer in Winterthur, Sulzer Brothers.

The first diesel engine sold for commercial purposes was a two-cylinder, 76-horsepower engine built at Augsburg and purchased by a Bavarian manufacturer of matches. Although the engine was installed in April 1898, it took several years to really iron the bugs out of that first machine, but it was reported to still be operating flawlessly 14 years after it left the factory. Now, the sale of license rights by Rudolf Diesel permitted him to begin establishing his fortune. Sulzer Brothers obtained the Swiss license for the engine as did the Belgian company, Carels Brothers. Others secured the rights in England, Russia, Denmark, Sweden (The New Atlas Company in Sweden had no relation to Atlas in California, as was sometimes suggested.), France and, because of the interest of brewer Adolphus Busch, the United States.

In 1898, when he moved to Munich to establish a research lab for continued development, Rudolf Diesel was well on his way to becoming a millionaire. He was 40 years old. He built a big villa of his own, one of Munich's finest. He also bought an N.A.G. touring car and, when he proved an inept driver, he hired a chauffeur. Despite his detractors, Diesel made an overwhelming impression on most who met him. Dr. Paul Rieppel, an engineer who worked for Diesel for a time and remained friends with him, recalled his first meeting with the inventor. "This slim man with his fine head and beautiful hands and his dignity and modesty gave no indication of his worldwide fame. Lively in his movements and concentrated in his talk, he exerted a suggestive force hard to resist. His balance of mind and body and his easy manner gave the impression of great strength."

Such impressions were typical. When George Dow of the Dow Pump and Diesel Engine Company in Alameda, California, met Diesel, he came away believing that Diesel was "the handsomest and most intelligent German that he had ever met."

But the good times were short-lived and, in fact, Diesel's success was always shadowed by financial and legal worries. A trip to America in 1912 to help Adolphus Busch lay the cornerstone of a new plant was only a respite. On his way home, he was lectured in Orange, New Jersey, by Thomas Alva Edison, who shared his opinions with Diesel about the perils of drink, the benefits of vegetarianism, and some peculiar ideas about internal combustion engines.

Diesel was fluent in several languages, including English and, during his 1912 visit, he granted an interview to *Motor Boating* magazine in which he briefly traced his engine's use afloat. According to Diesel himself, the very first marine diesel "was of 20 horsepower and was constructed in 1902-'03 in France for use on a canal boat by the French engineers, Adrian Bochet and Frederick Dyckhoff, in conjunction with myself. This engine had two pistons working in opposite directions and worked on the four-stroke cycle."

Diesel noted also the practicality of the two-cycle diesel engine "because scavenging is not done with the fuel air mixture but with pure air, so that not only untimely ignitions but also fuel losses are avoided." He described an experimental, 600 rpm, 30-horsepower, four-cylinder, two-cycle diesel that he felt would prove the diesel's practicality for small boats and yachts. He also estimated that, in 1912, there were about 365 diesel-powered vessels of which 140 were submarines, 40 naval craft, the remainder tugs or merchant vessels. He noted that the first American-built, diesel-powered ship was even then under construction. It was a 200-foot vessel for the Standard Oil Company and it was powered by a two-cycle engine built at the New London Ship & Engine Company (Nelseco), which had acquired a license from M.A.N. This engine had a 9 $\frac{1}{16}$-inch bore by 15 $\frac{3}{4}$-inch stroke and developed 300 horsepower at 300 rpm.

When Diesel returned home to Munich in 1912, his financial affairs did not become any easier to deal with. His spirits were low and his health was poor. Paul Rieppel was visited by Diesel at this time and recalled that Diesel "impressed me by his quietness and melancholy state of mind. He was filled by sad foreboding." Still, when his old friends Georges Carels and engineer Alfred Luckmann joined Diesel in

Antwerp on the cross-channel steamer *Dresden,* they had no idea of how low Diesel really was. The men were going to England to a stockholders' meeting of Carels' Consolidated Diesel Engine Manufacturers, Ltd. Carels and Luckmann had dinner with Diesel aboard the ferry, and the trio agreed to meet the following morning for breakfast. But Diesel never appeared. His stateroom, they found, had not been slept in and when they approached the ship's captain, they were presented with a neatly folded overcoat and hat found under the ship's rail. Rudolf Diesel's body was found two weeks later.

At the time of his death, Diesel's patents had recently expired. Despite some warnings from Adolphus Busch's American Diesel Engine company regarding its patents on fuel supply devices, the way was now open for development of diesel engines by quite a variety of companies on both sides of the Atlantic. In Europe, in February 1911 an ocean-going, diesel-powered ship had undergone her sea trials. Launched at Copenhagen, the *Selandia* was powered by two Burmeister & Wain diesel engines. These were four-cycle, air-injected engines that produced 1,250 horsepower apiece. But other sizable vessels with diesel engines were already operating and diesel-powered submarines had become very topical in Europe by 1903. All of this activity was watched with intense interest by potential diesel-engine builders in the United States. Big diesels were then being tested and the performance of the first diesel-engine ships was closely watched.

Although patent expirations may have prompted some companies to embark on diesel development, the whole field had been recognized as too promising to ignore by those who wanted to get in early. If Busch had been first to pay Diesel's license fee, he was not the only one to gain rights to European designs. Through a license arrangement in 1909, Electric Boat in Groton, Connecticut, used a number of Vickers diesels in E- and F-class submarines in 1910. These four- and six-cylinder, four-cycle engines were built for Electric Boat by Bethlehem Shipbuilding in Quincy, Massachusetts.

Recognizing that a significant market could develop from the Navy and another from commercial operators, Electric Boat then acquired the patent rights to the two-cycle diesel developed in Germany by M.A.N. A subsidiary company was founded by Electric Boat to build the engines in a newly erected factory in Groton. The New London Ship & Engine Company (Nelseco) began operation in the spring of 1911, and a commercial order soon followed. This was the engine to which Diesel had referred in his *Motor Boating* interview, a six-cylinder, 300-horsepower machine that weighed 32,000 pounds and was installed in a Standard Oil Company barge. This appears to have been the first installation of a commercial marine diesel in the United States, an engine that was not a big success. It was afflicted with problems with its two-stage air compressor, various other mechanical ills, and high oil consumption. The engine was removed a few years later.

It might be added that the two-cycle M.A.N. diesels were noisy. All these problems, together with the need to pay the license fee, seem to have been ample reason for Nelseco to abandon the two-cycle diesel. With the expiration of the Diesel patents, Nelseco's engineer, G. C. Davison, began work on a four-cycle engine. The engine was reasonably simple in its conception and featured horizontal overhead valves operated directly by big, vertical rocker arms that were themselves driven directly off the camshaft that ran along each side of the cylinders. This valve gear had proven itself to be exceptionally quiet in operation and was based on a system developed in England by Dr. Frederick Lanchester, among the most innovative of that country's

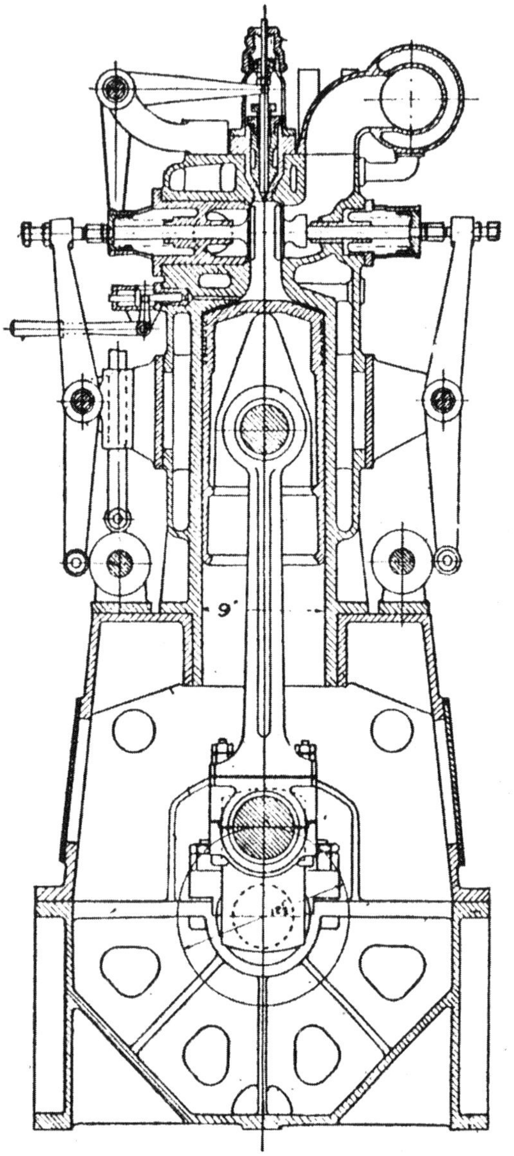

Nelseco diesel cross-section, 1913. This is the four-cylinder engine installed in the Alaskan cannery tender Warrior. *It had a 9" bore and 12½" stroke and developed a maximum of 125 hp at 400 rpm. Vertical rocker arms operated the horizontal valves. An air compressor was mounted ahead of the foremost cylinder. It maintained air pressure at 1,000 lbs. p.s.i. for use in starting the engine and to spray fuel into the cylinders.*

automobile pioneers. In a diesel, the horizontal valves proved to have another advantage. According to the company, the layout resulted in locating the fuel injection valve "quite a little distance from the piston top and [the valve] does not receive so much of the intense flame on one spot as otherwise would be the case."

The first of these new engines was a four-cylinder with a nine-inch bore and 12½-inch stroke that was rated at 120 horsepower. A Nelseco-patented fuel system with a single pump was used that was, according to the company, "made so durable and strong that there is little danger of disarrangement and no opportunity for trouble by operators changing adjustments, since there are no adjustments to change." When placed on the test stand, the prototype easily developed 115 horsepower and actually produced a maximum 137 horsepower, when pushed beyond its rated 350 rpm. A reversing gear was used rather than a direct-reversing mechanism.

The Nelseco must have been refreshing in its simplicity after the M.A.N.-style two cycles. The company reported that the engine's operation "is easily understood by men who were previously familiar only with gasoline and distillate engines, and is so very simple that there is little chance of its getting out of order in service. A demonstration engine was installed in an 84-foot motor yacht, the *Idealia,* and ads reported a fuel economy of six gallons of fuel oil per hour. "It would take 15 gallons of gasoline to produce the same power, *at approximately ten times the cost.*"

Perched on the edge of the New England coast, Nelseco had targeted its primary market even as the new engine was being laid out. The first engine sold, in fact, went exactly where the company intended, It was shipped to that region already buying semi-diesels in some numbers, the Pacific Northwest, where it went into the cannery tender *Warrior,* in 1913. The following year, a six-cylinder Nelseco powered what may have been the first U.S. diesel tug, the *Chicamauga.*

DIESEL
ENGINES
for
PLEASURE YACHTS
IDEALIA, 84′ x 14′ x 3′, is powered with a 120 H.P. N E L S E C O Diesel engine which drives her 14¼ miles per hour, using only six gallons of fuel oil per hour. It would take 15 gallons of gasoline to produce the same power, at approximately ten times the cost.
Everyone concedes the advantages of Diesel engines for commercial boats. The case of Idealia and other successful installations prove that N E L S E C O engines are equally suitable for substantial pleasure yachts and large cruisers.
The broad experience of this company, the first to build a Diesel Engine in America for marine use, is at the disposal of those who are interested in investigating or planning a Diesel powered boat. Our engineers will take pleasure in working out your power problem if you will put it up to them. 120 B.H.P. to 3400 I.H.P.—over 150,000 B.H.P in use or on order.
Let us send you complete information. Give us the details of your boat.
THE NEW LONDON
SHIP & ENGINE COMPANY
Groton, Conn., U. S. A.
Pacific Coast Office, 402 Exposition Bldg., San Francisco, Calif.
IDEALIA
The first American Diesel-powered yacht, and her NELSECO ENGINE.

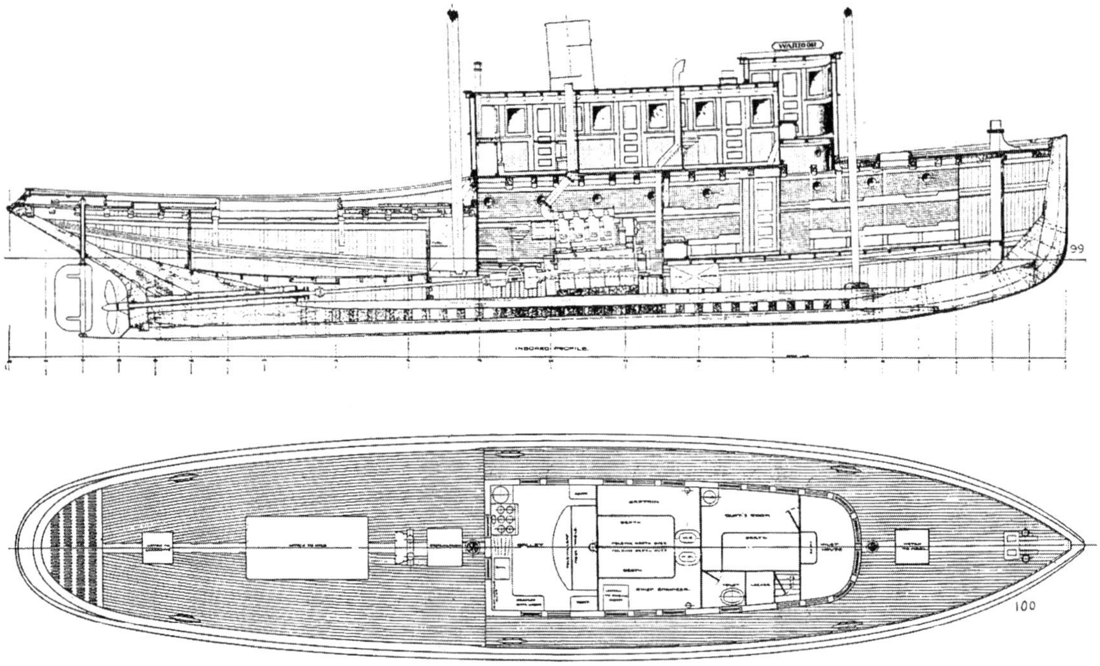

Warrior, *1913, first diesel-powered boat on the Pacific Coast. The 87-foot* Warrior *was built in Seattle by Nilson & Kelez for the Pacific American Fisheries Company. She was considered a luxurious vessel, with hot and cold running water and nicely finished fir sheathing in the engine room and crew's quarters. The engine room was entered by a watertight bulkhead at each end and housed the vessel's Nelseco diesel. A 12-hp Kermath stationary engine drove a generator for lighting, the air compressor, and pumps.* Warrior *was used in large part as a towboat but her fish tank could hold 18,000 red salmon.*

These big Nelsecos were joined in 1915 by a smaller displacement seven inch x nine inch, four-cylinder. It produced 60 horsepower at 450 rpm. In the period 1914–'19, Nelseco reported that it sold two dozen of its four-cycle engines including a pair of 60-horsepower versions, 14 120-horsepower models, and eight more producing 180 or 240 horsepower. Nelseco had now established itself and would eventually find its real niche as a builder of big, heavy-duty marine diesels that, by 1928, ranged in horsepower from 150 to 1,600 and in weight from 25,500 to 185,000 pounds.

The company was correct in its assessment of the sales potential represented by Oregon and Washington. In San Francisco and Oakland, California the men at Atlas and at Union recognized it too, just as surely as they understood that the end of the Diesel patents would sooner or later spell the end of the really big gasoline engine. Both companies would find healthy diesel sales to their north.

So did the Enterprise Engine Company, founded in San Francisco in 1886. Enterprise supplied castings to many Bay Area engine builders and built its own first gasoline engine, a single-cylinder, eight-horsepower model in 1917. This was followed by a pair of two-cylinder, 20-horsepower engines that had a 7 ½-inch bore and 9 ½-inch stroke and weighed 3,200 pounds. One of them was reported to still be in operation on a barge in 1955. Three years later, Enterprise introduced a diesel, a three-cylinder engine that developed 80 horsepower at 320 rpm.

The first Enterprise diesel had a Hvid-type fuel system patented by Rasmus Martin Hvid (Veed). A Danish-born, naturalized U.S. citizen, Hvid had established a company in Battle Creek, Michigan, to sell licenses for his combustion system, which was intended to improve vaporization of the fuel in the combustion chamber. The fuel was injected into a "cup" that communicated with the combustion chamber by a series of orifices. As the piston neared the top of its stroke, fuel in the cup ignited and this blew the entire charge within the cup into the combustion chamber. Reported *Diesel Progress,* "It was an attempt to duplicate the action of the injector air in the usual diesel engine of that time."

Hvid's design itself owed much to that of another inventor, a Dutch mechanical genius named Jans Brons, and Hvid acquired a Brons license himself to avoid conflict. Although many stationary engines would be built with the Hvid combustion system — including the St. Marys, the Sears, Roebuck "Thermoil" (built by Hercules in Evansville, Indiana) and a Cummins — Enterprise quickly turned away from the design. The Hvid system was sensitive to the quality of fuel oil, the cup's little orifices could plug with carbon, and it worked best at a fixed load and speed. Instead, in 1921, Enterprise obtained a license to use the common rail fuel system developed by Vickers engineer James McKechnie in 1915.

In 1924, Enterprise merged with the stationary-engine maker, the Western Machine Company of Los Angeles. Thereafter, the engine was known as the Western-Enterprise. These were highly regarded engines

Enterprise diesel, circa 1938

Army tugboat, circa 1942.

and the company introduced fully enclosed models that, by the early '30s, were using Bosch or Bendix fuel injection equipment. Walt Carruthers worked for 27 years for the Alaska Steamship Company and was on reasonably intimate terms with the whole generation of slow-turning diesel engines, including the Nordberg, Ingersoll-Rand, Busch-Sulzer, and others. Carruthers served as engineer on a 126-foot Army tug on a World War II-era trip from Washington to New York. The boat was powered by an 850-horsepower Enterprise that he remembers as "the most beautiful of all time. It had a chrome-plated valve cover for each cylinder and it was clean. A Fairbanks, by contrast, leaked at every joint and seam."

Given the Northwest commercial operator's potential appetite for semi-diesels and diesels, it would have been unusual if, in the industrial centers at Portland and Seattle, a local company didn't emerge to tap the market with a homegrown engine. In fact, there were two, and the one that got there first was started in about 1916 by Henry W. Sumner. As initially conceived and manufactured, Sumner's engine was a semi-diesel that required the combustion chambers to be heated by fuel oil fired torches. The engine itself was unusual, even by that era's standards. *Diesel Progress* described it as "an attempt to supply a simple marine oil engine which would especially appeal to those who felt that the running gear of any engine should be like that of the old, reciprocating steam engine. The operating engineer could see the gear in motion and could, if he wished, get his hands on it to be sure it was running cool. Such men did not believe in forced lubrication and closed crankcases."

Sumner's engine was a two-cycle with cylinders possessing a 16½-inch bore and 22-inch stroke. A four- and six-cylinder were offered, the four developing 350 horsepower at 210 rpm. The engine had an open base, "open column construction," for accessibility. The cylinders were supplied with scavenging air (the air necessary to cleanse the cylinder of burned gas and permit pure air to be compressed for combustion) via individual pumps. With the open base, provision had to be made for getting the air into the cylinders without loss of compression. This was accomplished by drawing the air into a sealed receiver that fed the cylinders. Main bearings did have pressure lubrication but others had wick-feed oilers.

Sumner found a market niche for these engines and, by 1926, was offering an 850-horsepower model, one of which was installed in a 176-foot ferry for the Sound Ferry Lines of Seattle. The following year, manufacture of the engines was taken over by the Willamette Iron and Steel Works of Portland. Willamette had been founded in 1865 and, by the mid-'20s, had grown to occupy a 12-acre site on Portland's waterfront. At the time that Willamette assumed production of the engine, the compression pressure was raised so the Sumner-Willamette operated as a pure diesel. As Willamette stated, "The engine starts cold without the aid of any igniting apparatus."

The elimination of "igniting apparatus" was the driving force behind the best known of the two diesels built in the Pacific Northwest. This was the engine developed by A. C. Estep, who was born in Sumas, Washington, and graduated from an "industrial school" in Aberdeen, South Dakota. He became what we might call a "practical engineer." When Estep returned to Seattle, he did so with money he earned from the operation and sale of a workboat on the Snake River in Idaho. Fascinated with engines, he secured a job with the Atlas-Imperial agent in Seattle and, at age 27, became responsible for installation of the big machines and related equipment. Shortly before the United States entered World War I, Estep and a lifelong friend named Fred Kimball took a 20-year lease on a prime waterfront property on the south side of Lake Washington's Ship Canal. There, they opened a repair shop named Estep and Kimball. Their timing proved excellent. There was plenty of repair work and this included some English submarines.

"Fred and I went on some of those submarines and looked the ships over. This is where I first got my ideas about diesel engines," Estep reported in an interview.

War's end revealed a new market to Estep. "After the war," he said, "there were many steam towboats whose engines were becoming obsolete, tired and in need of either extensive repair work such as new boilers, etc., or total replacement.... Ashes were all over the vessels, although the owners had little care for this, as long as the tug could pull a string of logs. Most of the tugs when fully loaded with water and coal were so deep in the water they did not have the necessary freeboard to go out in the ocean."

At the same time, the steam-powered fishing boats were facing similar problems of decay and inefficiency. Many were converted to big, gasoline engines, but engine-room fires were not unknown in boats powered by these engines. The Fairbanks-Morse CO, Bolinders, and other hot-bulb engines that operated on fuel oil began to grow in popularity.

"I decided I could improve on those early engines and make a profit," Estep said of his decision to develop a new diesel. He bought a heavy-duty gasoline engine and, together with Fred Kimball, used it as a basis for a prototype. He removed the cylinders with their 12-inch bores and replaced them with new nine-inch bore castings that would withstand diesel pressures. Then came cylinder heads, pistons, and most impressive, an Estep-designed-and- patented high-pressure fuel injector nozzle. Compressed air was used for starting.

"I had constructed a special flapper valve in the intake manifold which allowed compressed air to be introduced there, thence into the individual cylinders at the proper instant by means of the regular intake valves. This eliminated the need for separate air start valves in the cylinder head as on the Atlas. Once the engine began firing on its own, the compressed air was shut off, the flapper valve opened to the outside, and the engine continued to run."

Estep had this engine operating reasonably well rather quickly and it promptly drew the attention of Frank and John Frink whose father, John Sr., had given up schoolteaching in 1881 to found the Washington Iron Works. Initially begun in partnership with one L. H. Tenny, the Tenny Frink Company was reorganized in 1882 as the Washington Iron Works. The Iron Works was already a prosperous supplier of equipment to the lumber industry when the Frinks made an offer to buy Estep's patent and the rights to his engine design. Estep refused but joined the company as chief engineer, a position he held from 1921 to 1931 while also earning a dollar-per-horsepower royalty on each engine sold.

After he left the Iron Works, the engine, which had been known as the Washington-Estep was simply called the Washington. Estep planned to go back into business with his old friend, Fred Kimball, but when Kimball was killed in an accident, Estep moved to Two Rivers, Wisconsin, where, for a time during World War II, he was chief engineer at Kahlenberg Brothers. He later moved to Chicago and headed the Atlas office there before returning to Seattle to retire.

In the fall of 1922, the first Washington-Estep, a three-cylinder, was installed in the American Tugboat Company's *Elmore.* Six years later, the boat's chief engineer reported that the engine had "proved to be absolutely reliable under all conditions." He was then assigned to the tug *Irene,* equipped with a six-cylinder model. The Washington-Estep was a four-cycle diesel and, from the start, the company emphasized the advantages of common rail injection versus high pressure air: "Simplicity and reliability of operation. Fewer delicate parts with a reduction in maintenance cost. Greater ease in obtaining suitable operators. More reliable ignition when starting or at extreme low speeds due to elimination of the refrigerating effect of the expanding injection air; this also permits... lower compression and cylinder temperatures. Gain of 10 to 15% in mechanical efficiency due to not having to drive a high pressure air compressor . . . lower fuel consumption is obtained with perfect combustion and clear exhaust. Ability to burn any clean oil which will operate in any type of Diesel engine."

The metallurgy and manufacturing techniques used at the Iron Works suggests an engine of very high quality. The Works brewed up its own unique iron alloy for the base, which was then rough-machined, heat-treated, annealed and seasoned to eliminate any warping subsequent to final machining. The cylinders were fitted with heat-treated liners and had big water-jacket spaces. Valves were built of heat-treated iron heads and nickel-steel stems. Twin valve springs were fitted to ensure long-lasting, smooth operation. The valves were mounted in cages, the best practice in those times when a valve might need to be removed and replaced with some regularity. It could be done on the Washington-Estep without draining cooling water or removing the cylinder head. As delivered, each engine came with at least one spare intake and exhaust valve with cage. These impressive engines were built to order, each one requiring 60 to 75 days to manufacture for owners who, Estep said, had usually paid cash when the order was placed.

The Iron Works offered a wide range of engine ratings. By 1928, 33 different engines were cataloged. They ranged in horsepower from a 44-horsepower, two-cylinder to a 900-horsepower, eight-cylinder. The horsepower ratings, like those of most engines of that era, were conservative. In 1926, when a pair of 270-horsepower engines destined for two tugboats was tested, each proved to develop almost 400 horsepower.

Smaller models were equipped with the company's own reversing gear while larger models were direct-reversing, the big camshaft shifted in position by the use of 150 to 250 pounds of air pressure. "The engine can be started either ahead or astern in less than ten seconds. The governor control (or throttle) is adjusted for the desired revolutions wanted, while the maneuvering from ahead to astern is accomplished by a single hand wheel which slides the cam shaft longitudinally to bring the proper cams under their respective rollers, thus changing the timing to the right sequence and a slight pull on the air throttle starts the engine in motion."

An engine like this could be expected to make a lasting impression upon almost any inquisitive person of any age. Ed Larson was confronted by a Washington-Estep when he was 13 years old and he wrote vividly of the experience over half a century later in his book *Spring Tides.* He encountered a 180-horsepower Washington-Estep aboard the cannery tender *Doris E.* and remembered the engine like this. "I recall the huge injectors were about eighteen inches long and machined to perfection, sculptures in brass and steel. The injector lines were like hard, gold necklaces. The valve springs were black and massive, with silver push rods about four feet long contrasting beautifully against the forest green of the engine."

The fate of this wonderful diesel engine was typical but sad. The *Doris E.* was beached and caught fire at Petersburg, Alaska, in 1943 or '44, and her remains were later bulldozed out beyond the tideline, Washington-Estep and all. In one of those odd quirks of fate that occurs from time to time, an area resident saved the engine's 40-pound Johnson bar (the bar used, if necessary, to manually turn the heavy flywheel). Eventually, that mute reminder of a ship's life and an engine's life found its way to Ed Larson himself so that, as an elderly man, he now owns a poignant memento of a big part of his growing up.

As a matter of fact, these engines made an unforgettable impression on Estep himself. Looking back on his long career, he said that "the greatest pleasure of all occurred every time I went out on the trial trip when one of my new engines was put through its paces for the first time. What a thrill! When I went down in the engine room and looked along the line of cylinder heads, I could see all those rocker arms rocking forth and back as they opened and closed the valves with perfect timing. And then in the pilot house out in Puget Sound where we would plow up a white frothy furrow. Other vessels passed by, also powered with engines from our factory. It was wonderful to say to myself 'I designed those machines.'"

A slow-turning (200-300 rpm) diesel like this, as well built as was the Washington-Estep, had the same potential for long life as an Atlas-Imperial. It is estimated that at least 400 of the big machines were built. The most famous survivor replaced a steam engine in the 120-foot tugboat *Arthur Foss.* When launched in

Washington diesel tugboat Arthur Foss.

1889 for the Portland Railway and Navigation Company, the *Foss* was equipped with a 122-horsepower, double-cylinder, compound steam engine built in San Francisco by the Union Iron Works (no relation to the Union Gas Engine Company). In 1934, this engine was removed and a direct-reversing, 700-horsepower Washington-Estep was installed. It is there still, painted a creamy white and standing some 10 feet tall. It has a two-foot stroke and the volume of its six cylinders total 36,000 cubic inches. Even at idle, which is about 34 rpm when the engine is warm, the *Foss* will cruise along at 2 ½ or three knots.

On Christmas morning in 1929, 41-year-old Clessie Cummins drove through the streets of Columbus, Indiana, in a Packard sedan that had been repowered with a newly developed four-cylinder diesel engine. By then, Cummins had been working ceaselessly for about a decade to develop a low cost, light-weight, high-speed diesel engine. The Packard ran so well that he drove it to New York for the automobile show, burning a grand total of $1.38 worth of fuel on the trip. When the diesel-powered Packard was refused admittance to the car show, Cummins drove it to Detroit where it made a big hit at the meeting of the Society of Automotive Engineers. Thus began the development of the Cummins Model H. By early 1931, Cummins was using the engine to generate as much publicity as possible. One powered a stripped Duesenberg to a speed of 100 miles per hour at Daytona Beach. Another powered a truck on a cross-country trip at a fuel cost of $11.22. In December 1931, the truck circulated the Indianapolis Speedway for 14,600 miles without stopping. Cummins filed for a patent on the engine's distributor-type fuel pump and introduced additional models. By the time of the Boat Show in New York in 1934, Cummins was listing six models for marine application. The six-cylinder Model HMR developed 100 horsepower at 1,500 rpm and weighed 2,500 pounds, including its Joes reverse gear, a power-to-weight ratio of 25:1. Obviously, this was an engine of an entirely different conception from the big West Coast diesels. A 100-horsepower Washington-Estep weighed 18,500 pounds, a 185:1 ratio; a 100-horsepower Union diesel weighed 16,075 pounds.

"In whatever field we look," Clessie Cummins wrote in 1936, "we can see that the high-speed diesel engine has definitely entrenched itself as an accepted form of power.... The pioneering days we have gone through in the selling of the high-speed idea are beginning to bear fruit." For Cummins, the time when the "pioneering days" would "bear fruit" had been long in coming. In 1919, the company entered into an arrangement to supply Sears, Roebuck & Company with 4,500 1½-, three-, and six-horsepower engines. Most were returned to Sears by customers who either took advantage of the Sears 60-day guarantee or else because of problems with defects in the parts Cummins received from its suppliers. The contract was terminated in 1922 after some 3,000 engines had been built.

This episode was followed by problems with one-cylinder, Hvid-type diesels on the Gulf Coast in 1922. These difficulties included both carbon filling the holes in the injector cups and, even worse, mechanical failures. "In the next three years," wrote Clessie Cummins in his memoirs, "we designed, built and tested nearly three thousand separate fuel injection systems.... always our approach returned to the two basic ideas of eliminating the old separate fuel cup and using a plunger to drive the fuel charge out into the combustion chamber. Eventually, these two basics were synthesized in the Cummins injector which, along with the exclusive Cummins fuel metering system, has made Cummins engines outstanding in the high-speed diesel field."

With bad memories still fresh in the minds of Gulf Coast fishermen, Cummins took his next engine, the Model F, to Seattle in 1924. Despite its promise and even some later success with government agencies, the initial engines sold ran into problems that involved some poorly manufactured pistons, rings, and valves. What's more, when run at the low rpm needed for trolling, the engine's combustion chamber temperatures were too low to burn off the overly high asphalt content in the then-available fuel. Development of light grade diesel fuel would be necessary for the ultimate success of the high-speed diesel. But the refining of such fuel that, in part, grew out of the ongoing conferences of the fuel standardization committee at Penn State University would not occur until the mid-'30s. By 1939, with General Motors well along in its development of high-speed diesels, "light diesel fuel" became widely available.

As a result of the problems with the Cummins Model F in Seattle, the engine's only real customer proved to be agencies of the government. But additional difficulties developed when the engine was installed in power shovels, because, intended for marine use, it was not an enclosed design. That was rectified with the introduction of the "U" in 1928, which Clessie Cummins called the first completely enclosed diesel ever produced. Claiming a "first" remains risky business. In March 1927, for example, the Foos Gas Engine Company of Springfield, Ohio, had introduced its fully enclosed Type L diesel. This Foos was handsome in conception and in reality. It had a one-piece block with four valves per cylinder and solid fuel injection. Power-to-weight ratio ranged from 60:1 to just 30:1, depending on rating and rpm.

The Cummins Model "U" produced 10 horsepower per cylinder out of a 4½-inch bore and six-inch stroke and its compact size and enclosed design attracted great attraction at the 1928 Boat Show in New York. Although the first engines produced failed because of problems with poorly cast cylinder heads, the engine was successful after this problem was cured. The new engine had an electric starter, a major advance in making the diesel practical for smaller craft, especially pleasure boats. It also had the latest version of the

Single
Cylinder
7-8 H.P.
Bore 5″
Stroke 7½″

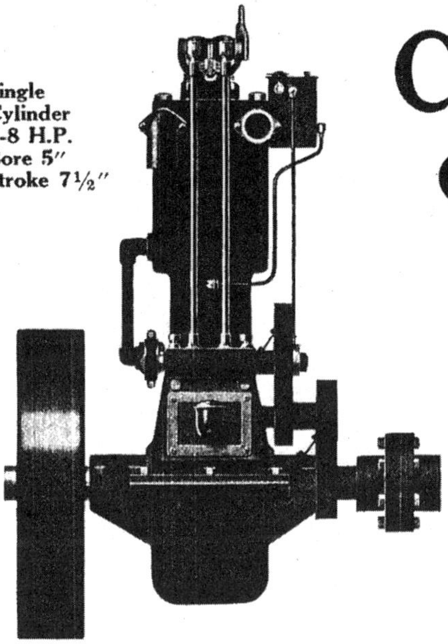

CUMMINS

Oil Engines

HVID TYPE

Burn Any Type of Low Grade Fuel That Flows Freely

Hvid Type engines are no longer in the experimental stages. Countless tests, both on the block and in actual service, have proved beyond doubt that this type of internal combustion engine is thoroughly practical, remarkably economical and absolutely reliable under every condition. To this type of engine is now added Cummins' ability—a combination that will be productive of engines of distinct merit.

Remarkable Economy and Freedom From Trouble Are Outstanding Features

Reliability, simplicity and economy—these three points have been held paramount considerations in the design of Cummins Oil Engines.

Every part is built to stand up under unusually severe conditions.

The design of the engine lends itself nicely to simplicity throughout. There is no carburetor or mixing valves—no electrical devices of any kind, such as magnetos, batteries, coils, wires, plugs.

As to operating costs, these engines are the most economical power plants ever designed. Any low grade fuel oil—kerosene, distillate and even most grades of crude oil—give perfect results and an abundance of power. Bear in mind that these engines are *designed throughout* to handle these fuels—not merely changed in parts. Their principle of operation is altogether different from that of the ordinary engine using gasoline.

These engines will run under a continual downpour of water—a feature that adapts them particularly well to the marine field.

Even in zero weather these engines will start immediately on any fuel oil that flows freely, ***without any preliminary warming up and without priming with some high grade fuel.***

From a long series of tests the operating cost of these engines has been determined as from one-sixth to one-tenth that of a gasoline engine. This feature, along with the feature of simplicity which permits the elimination of so many working parts, makes Cummins Oil Engines extremely desirable from every point of view.

Carbon troubles are unknown with these engines. Every particle of fuel is consumed—there is no residue. As no gasoline is required, there is greater safety and reduced insurance.

Back firing and kicking are impossible as the engine is not cranked against compression. Under sudden overload these engines will not slow down when the load comes on as do the common types of engines.

As the lubrication system is automatic and practically fool proof the chances of excessive wear are eliminated.

At present we are building and can make immediate delivery, on small quantity orders, two models—single cylinder, developing from 7 to 8 HP. on block test; double cylinder, developing from 14 to 16 HP. Bore and stroke of both models are the same—5″ x 7½″. Prices are well within reason.

Inquiries Solicited from Boat Builders, Marine Dealers, and Boat Owners

We urge an immediate investigation. We are establishing agencies in each boat center. If you can handle the sale of these engines in your territory write for complete details.

CUMMINS ENGINE COMPANY

Columbus, Indiana

Cummins Engine Company ad, 1921.

fuel injection system that Cummins engineers had been working on for so many years. Fuel was delivered by a plunger pump at a low pressure (30 to 50 pounds) to an injector in each cylinder. Suction created as the piston was on its downstroke drew fuel into a vaporizing chamber where it simply pooled above nozzles that led to the combustion chamber itself. In a 1928 *Motor Boating* article, Cummins described what happened next.

> "On the compression stroke, the full force of the charge of combustion air in cylinder, under maximum pressure of 500 pounds compression, which heats it up to 800 to 1000 degrees Fahrenheit, is forced through the small nozzles directly through the little pool of fuel, thus breaking it up into a hot, rich gas. This mixture is too rich to burn in the plunger chamber but, at top center, this plunger is suddenly depressed through a suitable cam mechanism, which drives this rich gaseous mixture out through the distributing nozzles into the red hot air in main cylinder combustion chamber. Here it mixes readily with the oxygen contained in the air and combustion takes place at once.
>
> "Because the charge is fully prepared beforehand, having been preheated and atomized through 720-degree crank travel, combustion takes place immediately on contact of the gas with the combustion air and, as a consequence, no advance in timing is necessary."

Cummins had made practical the intent of the Hvid patents that it had long since abandoned, and it produced an engine that offered control of operating speed that rivaled a gasoline engine. As introduced in 1928, the Model "U" was rated at 1,800 rpm. It was available in one- through six-cylinder models in ratings of eight to 48 horsepower. The latter weighed 2,750 pounds with its Joes reverse gear, a power-to-weight ratio of 67:1. By contrast, when Cummins showed its Model HMR in 1936, the 2,100 rpm, 55-horsepower version weighed 2,100 pounds, a 38:1 ratio. By then, Cummins was at last on the verge of financial stability. "In 1937," Clessie Cummins wrote, "after eighteen years of existence, the company operated with black ink on the books. My long battle had finally led to victory."

Cummins was not alone in its efforts to develop a compact, high-speed diesel engine that would serve for a wide variety of automotive purposes. In Harvey, Illinois, the Buda Company — named for the Illinois town in which it was first established in 1881 to manufacture railroad equipment — built its first gasoline engines to power its own track-inspection cars, and soon developed the line for automotive, marine, and industrial applications. Rising gasoline costs prompted Buda to consider manufacture of a diesel engine that could do all these things and it acquired rights to M.A.N. patents in 1926.

Then, Buda promptly introduced a four-cylinder, 1,000 rpm engine that, with six-inch bore and eight-inch stroke developed 92 horsepower. In 1929, a 2,000 rpm model appeared. The engines used an American Bosch fuel injection pump, a system that remains referred to as "solid injection." But its acquisition of the Lanova patent really established Buda. "Lanova" was a combination of the name Lang and the Latin word *Novum* – new. Franz Lang, one of the machinists who had worked with Rudolf Diesel at Augsburg, had developed diesel models of his own and focused on small engines whose efficiency was derived in large part from a "controlled turbulence" combustion chamber.

The benefits attributed to the Lanova combustion chamber were many, high among them the fact that the pressures developed within the cylinder did not exceed 625 pounds per square inch. This was relatively

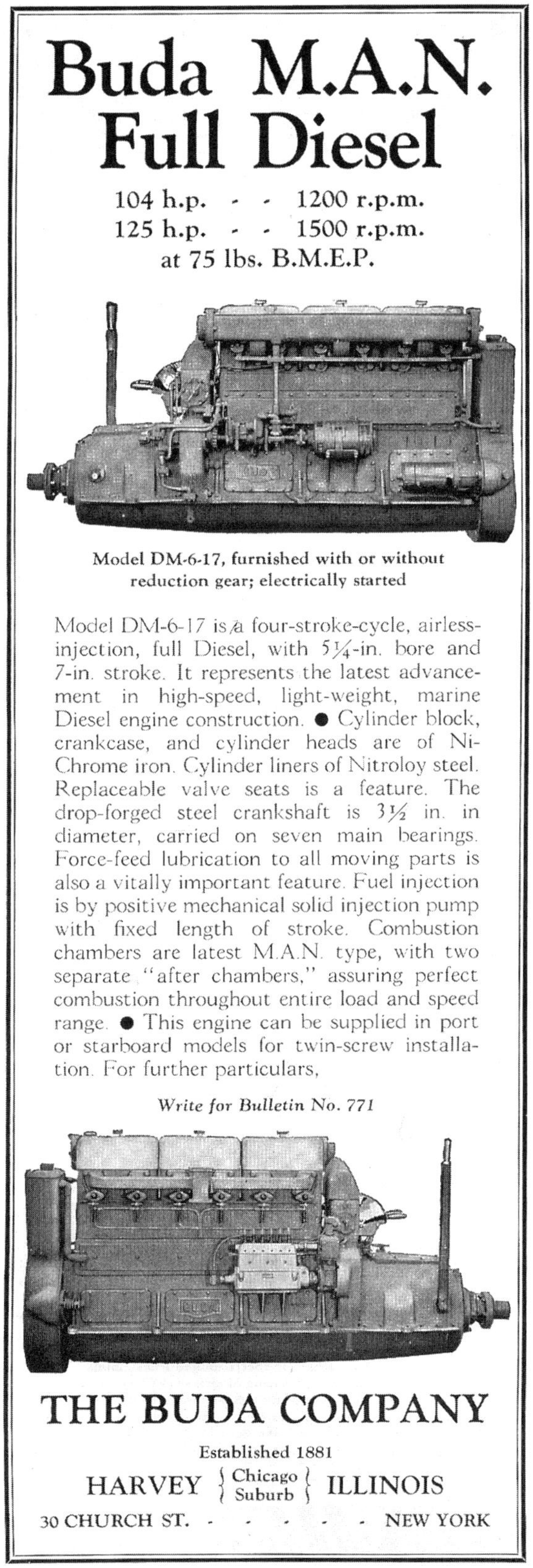

low and was the chief reason, the company pointed out, for the Buda's comparatively smooth, quiet operation. The compression ratio was relatively low as well, just 12:1 — glow plugs were required for starting — and the combustion chamber's shape was such that "no flame or fuel strikes the piston or cylinder wall. Consequently piston temperatures are no higher than is common in gasoline engines." The Buda marine diesels introduced in 1934 had power-to-weight ratios that varied from 22:1 to 35:1, including reverse gears. These Budas had so much appeal, and proved so reliable, that they soon attracted a very big customer, the U.S. Navy (see Chapter 5).

Throughout this period when Cummins, Buda, and General Motors, too, were at work on high-speed diesels, Atlas-Imperial continued to build the slow turning, cast-iron engines for which it had now become quite famous, engines with power-to-weight ratios of about 100:1. In the early '30s one of the original trio that had founded Atlas, Jack Lorimer, retired. This retirement, however, was short-lived. In 1936, Lorimer, together with his old associate Peter Bahr and Lorimer's son, Ralph, started a new company, the Lorimer Diesel Engine Company. Although these Lorimers were conceived as compact, lightweight engines, they retained the objective of simplicity so characteristic of the Atlas. But they employed a welded-steel crankcase and were fully enclosed. Some smaller models were designed to start on gasoline and then be switched over to diesel operation. This, however, caused problems with insurance and the company was dissolved shortly after World War II, at which point one of its engineers, Roy Anderson, became chief engineer of the Winslow Engineering Company.

Atlas-Imperial entered the Depression with as great a challenge as any other manufacturer. Wesley Webster, an engineer at the old company, found himself and his colleagues working double shifts and, as necessary, sweeping the factory floors. By the mid-'30s, Atlas itself

Lorimer diesel, circa 1944. This eight-cylinder engine had a lightweight, welded-steel crankcase.

Lorimer diesel, circa 1944. This two-cylinder Lorimer diesel was started on gasoline, and the carburetor can be clearly seen in this photograph.

Atlas diesel, circa 1940. This is the fully enclosed "medium-speed" Atlas that was rated at 900 rpm.

was looking for ways to extend its product line and to offer engines less expensive than, for example, a 60-horsepower engine that sold for some $6,000, the cost of a good, four-bedroom home. A medium-speed (900 rpm), enclosed model with bore/stroke dimensions of 6¼" x 8¼" was introduced in three-, four-, and six-cylinder models. The latter developed 135 hp and, like the others, was available with reduction gears.

In January 1936, Atlas introduced a new, compact, high-speed model named for its fuel system, the Atlas "Atimco." The engine's key features were set forth as follows:

> "Aside from the reduction in weight and general refinement in design, in keeping with modern Diesel engine practice, the outstanding feature of the engines is the fuel system.... It is a development of the common rail, constant pressure system that has been a distinguishing feature of the Atlas-Imperial Diesel since its inception.
>
> "In the new Atimco system, the desirable features of the common rail have been retained, but the delicate adjustments required to equalize load between cylinders have been eliminated. A single fuel pump maintains constant pressure in an accumulator which is in communication with a distributor and fuel metering valve.... Since the same metering valve measures the oil for each cylinder, there can be no variation in the quantity of fuel injected into each cylinder, and no unbalanced load on the engine."

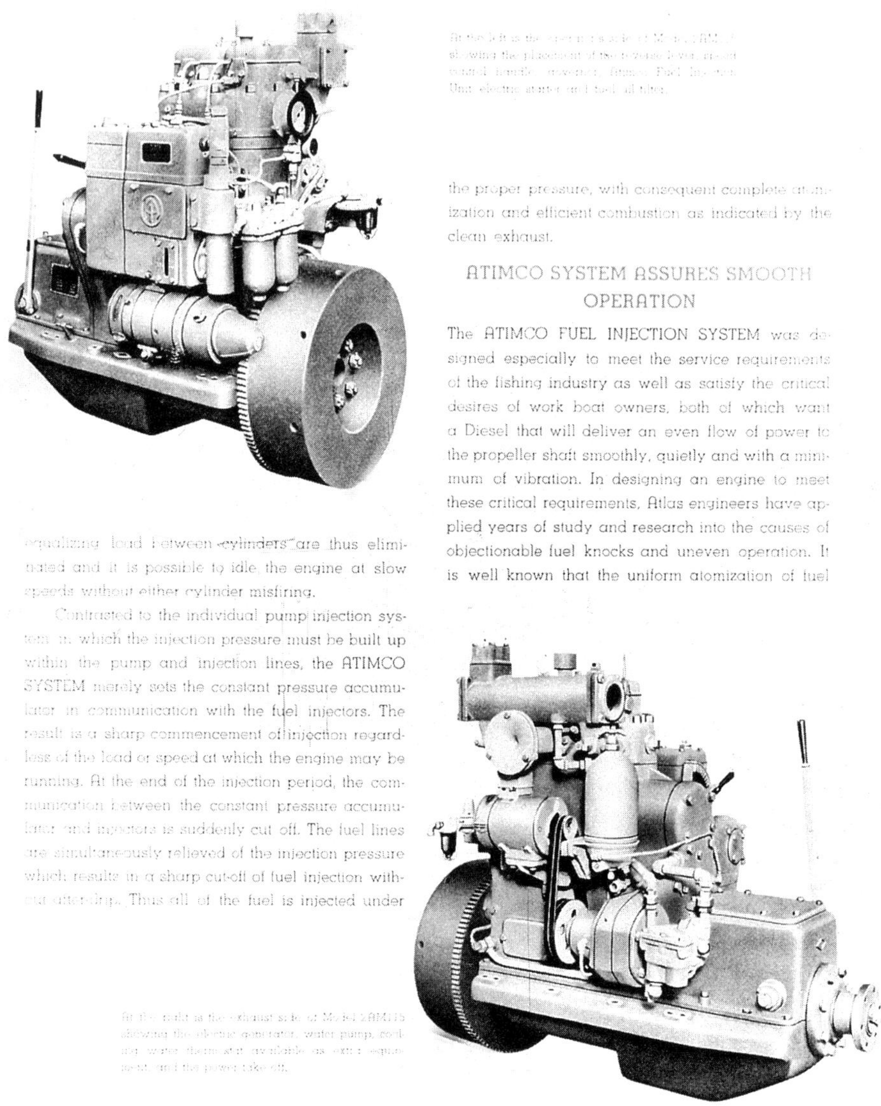

the proper pressure, with consequent complete atomization and efficient combustion as indicated by the clean exhaust.

ATIMCO SYSTEM ASSURES SMOOTH OPERATION

The ATIMCO FUEL INJECTION SYSTEM was designed especially to meet the service requirements of the fishing industry as well as satisfy the critical desires of work boat owners, both of which want a Diesel that will deliver an even flow of power to the propeller shaft smoothly, quietly and with a minimum of vibration. In designing an engine to meet these critical requirements, Atlas engineers have applied years of study and research into the causes of objectionable fuel knocks and uneven operation. It is well known that the uniform atomization of fuel

equalizing load between cylinders are thus eliminated and it is possible to idle the engine at slow speeds without either cylinder misfiring.

Contrasted to the individual pump injection system in which the injection pressure must be built up within the pump and injection lines, the ATIMCO SYSTEM merely sets the constant pressure accumulator in communication with the fuel injectors. The result is a sharp commencement of injection regardless of the load or speed at which the engine may be running. At the end of the injection period, the communication between the constant pressure accumulator and injectors is suddenly cut off. The fuel lines are simultaneously relieved of the injection pressure which results in a sharp cut-off of fuel injection without afterdrip. Thus all of the fuel is injected under

Atlas Model 2AM115, 1936. This two-cylinder engine developed 18 hp at 950 rpm It was an effort by Atlas to bring diesel power to small fishing and working craft. Its newly developed "Atimco" fuel system was based on a constant-pressure accumulator that resulted in notably even and vibration-free operation. Said the factory literature: "Contrasted to the individual pump injection system, in which the injection pressure must be built up within the pump and injection lines, the ATIMCO SYSTEM merely sets the constant pressure accumulator in communication with the fuel injectors. The result is a sharp commencement of injection regardless of the load or speed at which the engine may be running. At the end of the injection period, the communication between the constant pressure accumulator and the injectors is suddenly cut off. The fuel lines are simultaneously relieved of the injection pressure, which results in a sharp cutoff of fuel injection without afterdrip. Thus all of the fuel is injected under the proper pressure, with consequent complete atomization, and efficient combustion as indicated by the clean exhaust."

Atlas was seeking to develop a compact, high speed, diesel engine at the same time as Cummins, Buda, and GM. Whatever the merits of the Atlas system, it did not achieve market success.

Crankcase with crankshaft, Atlas-Imperial factory, Mattoon, Illinois, circa 1940. (Courtesy, Dan Webster)

Atlas intended the Atimco for the owner of the smaller fishing or workboat seeking the durability and fuel efficiency that was associated with its older models. Electric starting was offered as were an optional generator and power takeoff. The company embarked on an ambitious manufacturing effort for the new model. In 1935, it purchased a factory at Mattoon, Illinois, and this plant was in full production by January on the new-style Atlas as well as the older version. The following year, Wesley Webster was transferred from Oakland to Mattoon to be the works manager and engineer.

The Atlas Atimco was not the company's only effort in the market for more compact, higher-speed diesels. In 1938, Atlas announced that it would distribute "light weight, high-speed" Buda marine engines. One ad included the logo, "At Present Prices, Boat owners can easily afford a Buda Diesel Engine." The company also introduced an enclosed, heavy-duty model, and a medium-speed (900 rpm) fully enclosed version. Today, Atlas specialists have little knowledge of these engines, the Atimco, or the enclosed, medium-speed engines, which appear to have made no lasting, positive impression.

Instead, the big news always tended to center around the classic Atlas-Imperial. In 1939, it was reported that the "company played an important part in stepping away from old, established habits" by powering the first diesel-electric tuna boat ever built. This vessel, the *Challenger,* was soon taken into the Navy and spent the war in the South Pacific. Even at the close of the war, Atlas was trying to find a way to retain its slow-turning engine but get more power out of it. In 1945, Atlas offered the option of a turbocharger designed by a Swiss engineer, Dr. Alfred J. Buchi. Cooper-Bessemer had introduced this turbocharger in 1935 and a year later the American Locomotive Company used it on one of its diesels. So equipped, a big Atlas could churn out up to 900 horsepower. But this was not enough to save the company in the postwar economy. In fact, nothing was enough.

"What happened to Atlas," remembered John Seymour, a lifelong diesel specialist in Oakland, "was that it fell apart because it couldn't keep up with the times. Around 1950, White-Superior bought the company to get the name and the rights to make spare parts."

Back in Illinois, during the war, the Mattoon plant produced engines for the government, probably generator sets. The factory also manufactured 155-millimeter artillery shells and it was recognized for its contributions to the war effort in a letter from the government. But with the postwar sale of the company, the Mattoon factory was purchased by a manufacturer of radiators, and the engine staff was dispersed. Wesley Webster moved up to Chicago and went to work for Venn-Severin, an old-time company founded in 1906 by Theodore Venn and F. W. Severin to build oil engines and, in 1920, its first diesel.

In America, the end of World War II would mark the beginning of a new age, social, political, and technological. It would also mark the end of much that was old and familiar. In the world of commercial vessels, the first thing to go was the slow-turning, cast-iron diesel engine that had served a whole generation. The seeds of its demise were planted well

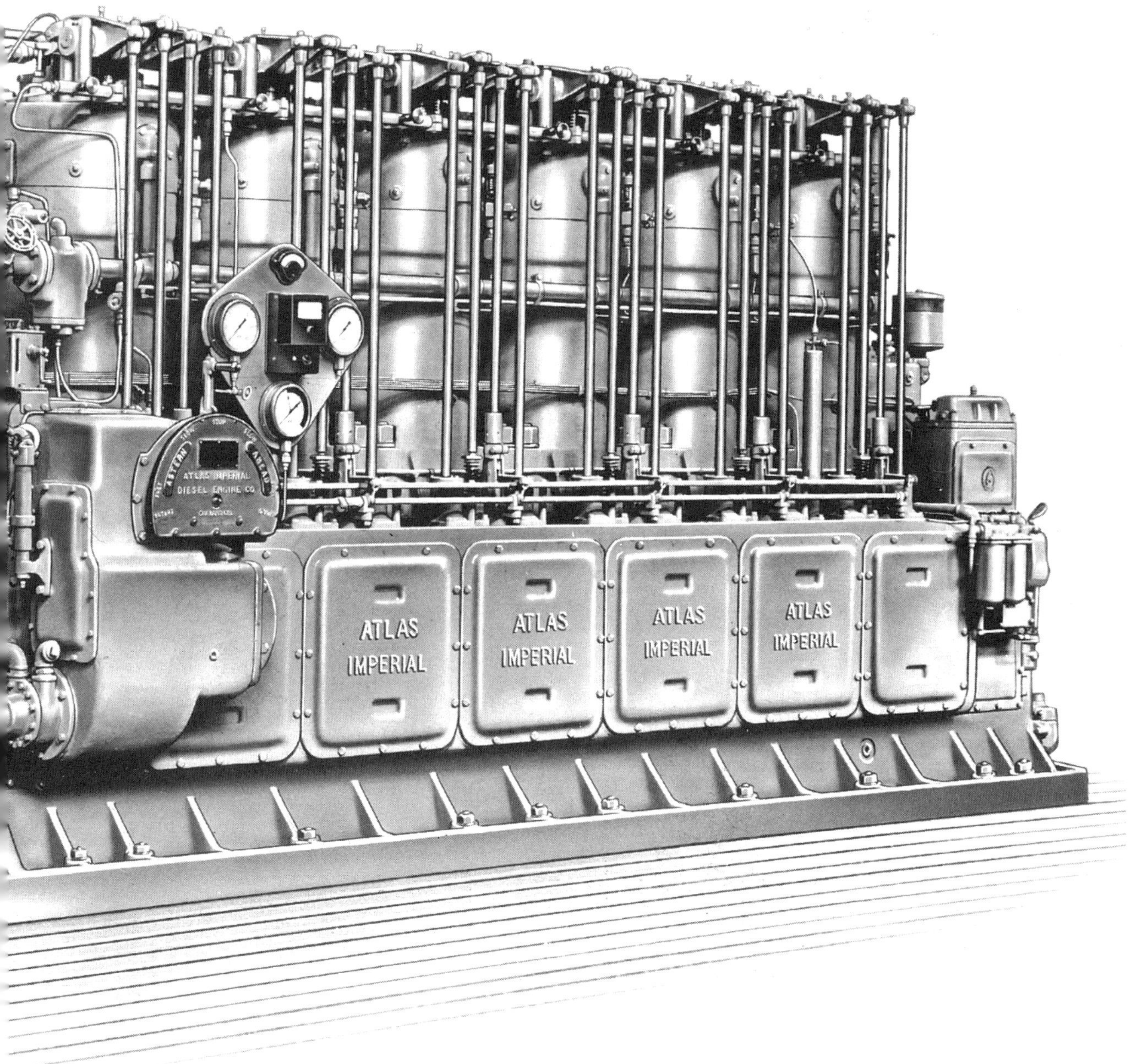

"Turbo supercharged" Atlas Diesel, circa 1945. Here is the last of the big, traditional, Atlas-Imperial diesels. Engines like this were still being installed after World War II and the Lucy Elena, *a steel San Diego tuna clipper had a 550- hp Atlas installed in 1948.*

before the war in the efforts of different companies to improve its efficiency, affordability, and most of all, its power-to-weight ratio. But the end of war also brought the unforeseen availability of thousands of war-surplus "Jimmys." What is more, on the Gulf Coast, a totally unexpected new burden was added to the already problematic future of the old-style engines. Offshore oil rig companies became willing to pay boat operators based on the horsepower of their vessels. A 165-horsepower Series-71 diesel immediately meant more money could be made than with an 80- or 100-horsepower Atlas-Imperial. What's more, the Jimmy could be had brand new for a few hundred dollars. All along the Gulf Coast, the old Atlas diesels began to be removed.

At first, to old-time diesel men, to engineers like Walt Carruthers who had been shipmate with all manner of big, marine diesels, the new engines seemed almost like a joke. "They were noisy and we hated 'em," he remembered. "An old engine, you ran by feel. The new ones would just shut off if something went wrong. We laughed at the Series 71 and called them sewing machines. But the fact was, there was more horsepower per cubic inch in one of those engines than any diesel in the world."

Gas boat Crane, *Little Kanawha River, West Virginia. It's tough going for the Model T driver who needs added horsepower to haul the hay that appears to have just been off-loaded from the* Crane. *Small stern-wheelers like this were a staple on the rivers and ruled the Little Kanawha with its shallow-water bars and tight bends. (Courtesy, Jack Custer)*

Chapter 2

Rivers

"A great deal of commercial work has been accomplished on the rivers with boats of 15 and 20 horsepower.... The raucous starling flock of gasboats infiltrated the rivers."

—"Remembering the Gasboat," Fred Way, Jr., *S&D Reflector,* March 1996

"The application of Diesel power was not confined to use in yachts and railcars. The first Winton Diesel installation in a tugboat was the tug D.M. Benton *in the year 1923. In the same year, the first Diesel-electric drive was installed in the towboat* J.B. Battle. *During the years from 1923 to the mid-'30s, it became evident to operators of workboats that the development of the Diesel engine had reached the stage where it was unquestionably the power for this type of service."*

—from an untitled internal history of the Winton Engine Company and Cleveland Diesel Engine Division of General Motors, circa 1962

Because it forms a triangle bounded on its north and south sides by two rivers, Pittsburgh, Pennsylvania, has always been known as a city of bridges. In the days when they bore high volumes of pedestrians, these bridges served as wonderful places to watch the river traffic. Steam-powered towboats of all sizes made their way along the Allegheny and the Monongahela and then, right at the tip of the triangle, the stern-wheelers entered the mighty Ohio. It was, in the early years of the century, a city of smoke. Fires from the blast furnaces of the riverside steel mills colored the night-time rivers wine-red with their reflections, and smoke from the stacks darkened the sky. Even the city's leading steel maker, Andrew Carnegie, was concerned about the smoke. He predicted great rewards for the man who would rid Pittsburgh of its dirty air and soot. On the three rivers, the ever-busy steamboats added their share of smoke to the all-encompassing haze.

Then, one early spring day in April, 1907, pedestrians on three of Pittsburgh's Allegheny River bridges, the ones at Sixth, Seventh, and Ninth streets were stopped in their tracks by the sight of a new sort of boat. Said the *Pittsburgh Press:* "The first motor towboat seen in Pittsburgh waters chugged up the Allegheny river yesterday afternoon with a small flat [barge]. It is owned by a junk peddler who has foresight."

For years, the owner of this boat had been collecting and selling his wares in the little mill towns all along the three rivers and one day, apparently, he got tired of rowing against the current. "A few days ago, he invested in the motor boat and the flat, and now does business at top speed," is how the paper put it. But what so surprised the onlookers on the bridges that day was the sight of a boat moving up the Allegheny River at a good speed, with only one man aboard, and... "*without smoke*."

By the turn of the century, the recognition that gasoline engines could ease the burden of labor had become evident to hardworking individuals all along the country's vast river systems. One didn't need to be working on a river steamboat to perceive that a new era had arrived. Down the Ohio, well south of Pittsburgh at Point Pleasant, West Virginia, a ferryman named John Henry Stone reckoned that a gasoline engine could replace the hand-line system he used to haul his little flatboat across the Kanawha River to Henderson. Today, Charlie Stone doesn't know the exact year when his grandfather first motorized his little ferry, but guesses that it was sometime in the very early 1900s. He remembers that by the time his father began work on the river as a boy in 1902 or '03, small, gasoline-powered push boats, locally referred to as "yachts" were already common.

"He bought the engine in Cincinnati," said Charlie, "probably sometime in the late 1890s. We're not quite sure of the year or the make of engine. But he did need the engine because traffic was growing. The yacht was tied alongside the ferry flat." The flat itself, built of white oak, was large enough to carry a team of horses on its plank deck. John Henry Stone's yacht was similar to a blunt-bowed john boat and about 12 feet long. Often, these boats were unpainted, but Stone painted his white and black.

There were men like John Henry Stone on the rivers from the Gulf Coast to the Dakotas, from the Atchafalaya and the Mississippi to the far reaches of the Missouri, from the Beaver and Kanahaw to the Snake and the Yellowstone and the Columbia. In 1906, for example, the *Boise News* reported that a "gasoline steamboat of a type entirely new to northwestern waters" would be used as a supply boat in the lower section of the Snake River Canyon. The 36-foot boat had a pair of 12-horsepower motors and a carrying capacity of 10 tons. In 1910, the United States Reclamation Service began using a 20-foot launch powered by a seven-horsepower, two-cycle Ferro engine built in Cleveland to ferry a barge used in a construction project on the Salt River at Roosevelt, Arizona. A year later, at Poplar, a Montana paddle-wheel ferry driven by a two-cylinder, 24-horsepower Gray motor was built by James Avery to carry passengers and freight across the Missouri.

While, in many cases, the internal combustion engine replaced manual propulsion on the inland rivers, or four-legged horsepower on the old Erie canal, it made its most visible impact as an alternative to steamboats. The rules that had for so long regulated the operation of steamboats were now no longer applicable. Said *The Gas Engine* magazine: "The Government regulations allow one or two men without long apprenticeship or special training to man a boat propelled by gasoline, while the steamboat is compelled to carry a much larger crew, and part of them must be licensed and stand a severe examination."

Neither did the "gas boats," as they were called, need to carry laborers to pass coal from riverbank bunkers or a coaling flat lashed alongside the steamboat convenient to the boiler room. The ability to operate with unskilled help and comparatively small crews, and the lower cost of fuel (In the Ohio River valley in 1911, gasoline sold for 12 cents per gallon; kerosene cost nine cents.) and maintenance gave gasoline-powered boats an immediate advantage over steam that would only be increased with the advent of semi-diesel and diesel-powered vessels. The smaller steamboat, in particular, was about to become economically obsolete.

Today, the sons and grandsons of the pioneer rivermen who built and operated the first gas boats all remember 1900 as a general date that marked the introduction of these craft. By then, in fact, such boats were probably seen as new, but not entirely novel. And just a decade later, on the Ohio and Mississippi and the smaller streams that fed them, thousands of gasoline engine stern wheelers were reported to be in operation towing, carrying lumber, sand, freight and passengers, and being used in herring and clam fisheries.

A journalist of the epoch, close enough to the beginning of the gas boat era to be credible, attributes its beginning to a period around 1891 and an unknown genius who perceived that the stationary, horizontal engines first introduced could almost have been designed for adaptation to a stern-wheel riverboat. This appears to be as good a guess as any regarding dates, although a stern-wheeler on the West Coast may have beaten everyone else. In 1890, at Scotia, California, W. C. Harris installed a 15-horsepower engine, almost certainly a Regan, in a boat named *Ruth,* which he operated on the Eel River. This event made big news in Humboldt County, and was acted upon at once by the most enterprising launch operators down in San Francisco Bay. Among the earliest of the eastern boats for which there survives a record is *Gipsy,* a 44'6" x 8'4" scow owned in Marietta, Ohio, by the Ohio River Sand Company. According to the company, the boat's 10-horsepower engine was built in 1891.

With the stationary, single-cylinder engine, a pulley or pulleys affixed to one end of the crankshaft drove a pair of long leather belts, eight- or 10-inches wide, that, at their aft end, drove a countershaft. It, in turn, drove the paddle wheel via sprockets and a chain. The massive cylinder of such engines and the ingenious drive arrangement made a lasting impression on Sewickley, Pennsylvania, riverboat chronicler, Fred Way, Jr. "The belts reached aft," remembered Way in an article on gas boats in the *S&D Reflector,* "one crossed and the other straight to a common jackshaft rigged with more pulleys and belt-shifting gear. The jackshaft on one end or the other (sometimes both) had a toothed sprocket which, by link chain, was connected to a much bigger sprocket and the paddle wheel shaft. There were variations on this theme but not often. Sometimes one belt served, in which case the reversing was done at the jackshaft."

Gas boat at Prairie du Chien, Idaho. It is rare to find a photograph showing the engine installation in a gas boat. Here we see the twin flywheels of a stationary engine, the port-side wheel rigged with a pulley for the drive belt. (Courtesy, Jack Custer)

Gas boat, unknown western river. Probably used as a ferry, this little boat has the drive system that was typical of most such craft. A canvas or rubber belt from the engine drove a jackshaft which, in turn, drove the stern-wheel by means of a chain. (Courtesy, Jack Custer)

Experience proved to some that a malleable iron chain was superior to steel, which was likely to snap in freezing weather. The right kind of steel roller chain, however, was said to last for years. Larger boats were equipped with a "chain catcher" made of strap iron that was mounted under the chain. The system of pulleys and countershaft formed a natural reduction gear so that an engine running at 200 rpm might be turning the stern wheel at 35 to 50 rpm. One of the two drive belts was twisted so that, when engaged, the boat would operate in reverse.

Such boats were, by 1910, if not earlier, a fixture from Pittsburgh to New Orleans. There, 50- to 80-foot paddle-wheelers that drew only 14 inches and were driven by big stationary engines were considered the region's most important powered boats. "Their business," reported *MotorBoat* in 1913, "takes them into the shallowest streams and bayous which are little more than mudflats covered with water, either from the gulf or from the river.... They carry back to the small villages, to the plantations, the merchandise which the city furnishes to the men and women who dwell in these out-of-the-way places. Their work is always in quiet and shallow waters....Of these there are about 100 in New Orleans waters."

During the first 10 or 15 years of gas boat development, there were apparently numerous efforts to improve upon the single-cylinder, belt-and-countershaft drive system. Higher speed, multi-cylinder engines were most typical of these efforts, driving the stern wheel by a shaft and transmission. It was a method that required very precise alignment of the shaft and then suffered from gear binding as the boat rolled. Then, too, the starting torque of the multi-cylinder engine was found wanting. The big, horizontal-engine, with its heavy flywheels, had proven itself to provide immediate torque and its long stroke was well adapted to keeping the boat moving ahead. Rivermen had proven to themselves that such engines had more towing "grit" than a multi-cylinder of the same horsepower. (Note that the word "tow" has always been a misnomer on the rivers as the boats always pushed — "pooshed" in the vernacular of rivermen whether they come from the North or South — their barges.)

Individuals and small boat yards were turning out vessels of the *Gipsy*'s general type well before the turn-of-the-century. Riverman H. E. Robinson built *three* boats between 1898 and 1910. His 62' x 14' x 3½' *Lillian* may be considered typical of the type. The boat had a 20-horsepower stationary engine with a straight belt and a cross belt. On an extension shaft of the engine, Robinson mounted a 24" x 30" pulley on which the drive belts rested. Said Robinson: "I find it pays to build a gas boat on steamboat lines, and when so built they are on an equal with boats of far larger proportions driven by steam." The *Lillian* proved able to tow 200 tons against a stiff current, a performance that surprised at least the steamboat operators who witnessed it.

The chief objection to early one-cylinder stationary engines for marine applications appears to have been related mostly to vibration. Fred Way, Jr., who did his best to annotate details of every boat on the inland river system, remembered the early gas boats from his childhood days on the Ohio. He always recalled that, with the boat's big one-lunger thumping away, "the decks were jumping up and down, the boat shaking violently, fuel drums on the forecastle in a continual earthquake and the big, straw-hatted pilot up there in his pilothouse bobbing his head and belly in rhythm to the trembling, wooden steed under him."

In part, the vibration was caused by the method of throttling these stationary, "hit-or-miss" engines. When such an engine reached its intended rpm, the governor moved a rod that permitted the intake valve to remain closed, and the exhaust open so that the engine "missed." The flywheels then kept the machine turning until, when the speed slowed to desired rpm, the governor again permitted the rod to release the valve gear.

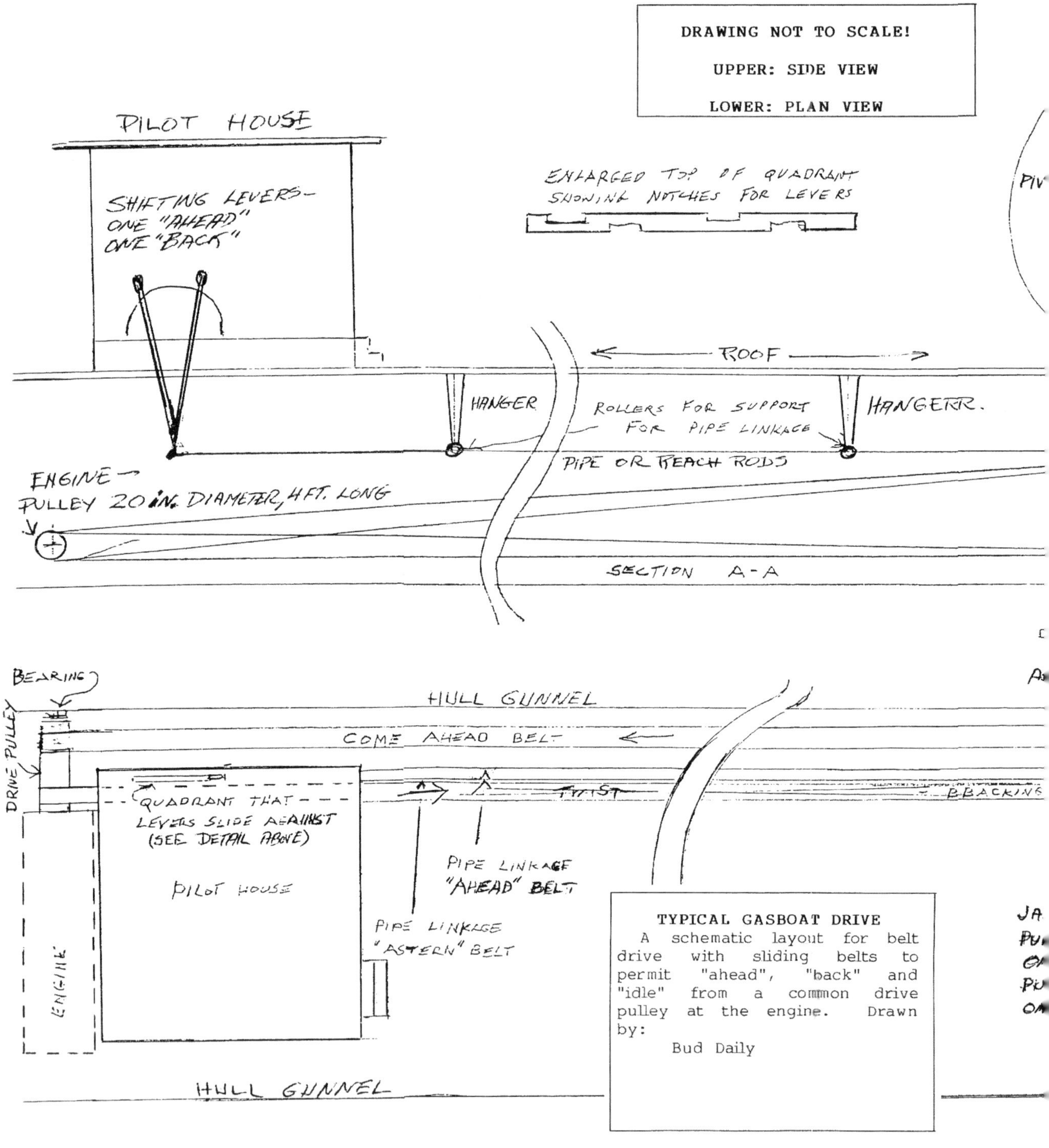

Gas boat drive system. The profile and plan view, sketched by riverman Bud Daily, show the layout from pilothouse to stern of the belts, pulleys, rods, shafts, and chains used to transmit power from the gasoline engine to the stern wheel. (Courtesy, Bud Daily)

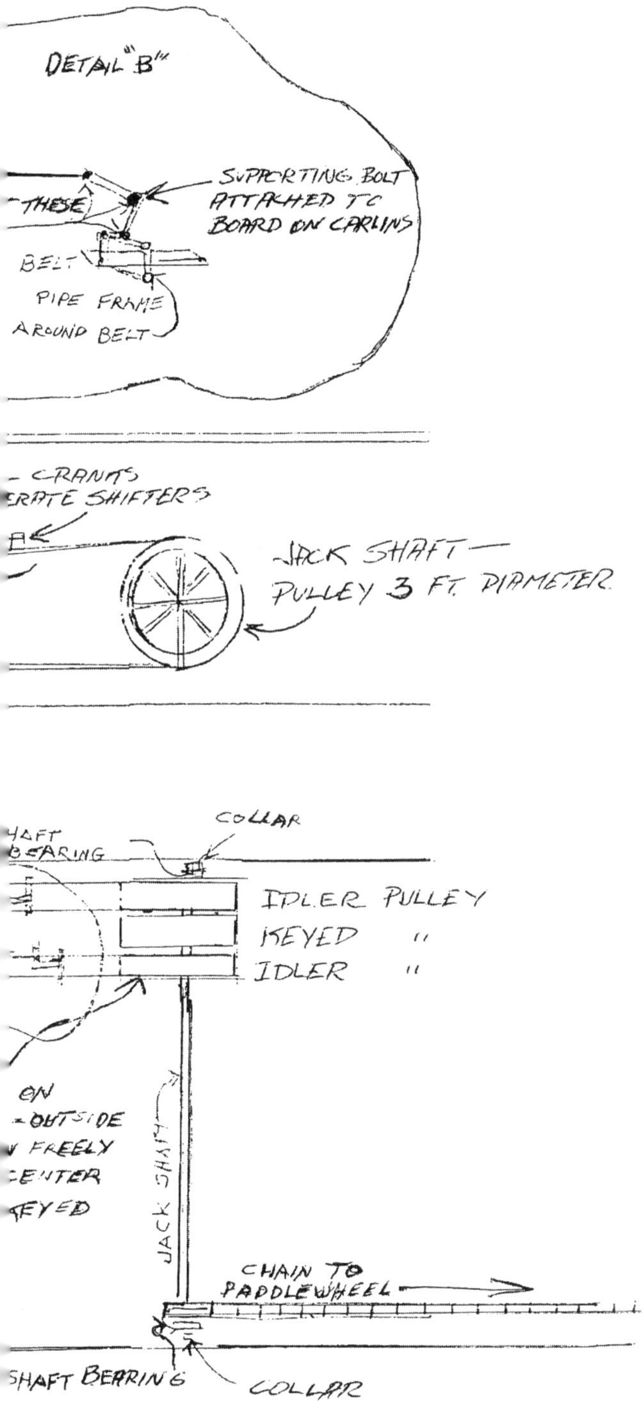

The substitution of a "throttling governor" for the one-speed, hit-or-miss-style governor supplied with the engines helped in some instances, but vibration was an issue that challenged single-cylinder proponents for years.

"We had difficulty in convincing [operators] that our single-cylinder engine was better than any multiple-cylinder," said the Midwest agent of the Otto Engine Works in 1911. "It took years to prove it to them, for, like a great many people, the Upper Missouri River boatmen first insisted on trying an inferior grade of engine because it was cheaper in the first cost. . . . They first equipped [one boat] with a 60 hp single cylinder engine, which was a failure, and thinking they did not have power enough, they put on an 80 hp twin engine of another make, which likewise was a failure, and they finally consented to allow us to install one of our 60 hp engines and immediate machinery, which was entirely satisfactory and which machinery is still driving the boat after twelve years." This boat was the 94'x 18' *Iowa,* berthed at Chamberlain, South Dakota.

Business for the Philadelphia-based Otto on the upper Missouri was apparently good in the years leading up to 1900 and during the first decade of the new century. Sales agent Marshall reported installation of "scores" of engines in the 15- to 80-horsepower range. A 43-horsepower engine was installed in a 120-foot stern-wheeler named the *John Youngblood* in 1896 and the boat was still running as the *Bismarck* 15 years later. This was but one of many instances of the comparatively long service lives for gas, and later semi-diesel and diesel boats by comparison with steamboats. Another Otto-powered boat, the *City of St. Pierre,* was equipped with an 80-horsepower engine in 1904 and was known as the fastest boat on the Missouri.

These comparatively large boats had a somewhat different drive system than smaller river craft. Now, the drive belt ran from the engine beneath the deck to a jackshaft. It drove the paddle wheel via spur gears and sprocket wheels through a duplex clutch. Power was carried to the paddle wheel through roller chains. Pilot house controls included a tiller to control the brake on the jackshaft, another to control the clutches, and a throttle. Said Marshall: [The pilot] "has much better control of the boat than is possible on a steamboat, where the signals have to be communicated to an engineer and then acted upon by him with the possibility of a misunderstanding of the signals."

Although Otto engines had a dominant role in powering the early boats on the upper Missouri, the brands of engines used elsewhere in the early days often went unrecorded. In part, this was caused by a refusal of some of the day's magazines to publish the engine makers' names in order to deny them free publicity. One writer in 1911 estimated that 80 percent of the boats on the Ohio between Pittsburgh and Cincinnati were equipped with one maker's engine, but he was not permitted to name the company. (It was most probably Fairbanks-Morse.) What's more, engines quickly became so taken for granted that a boat's power was often referred to as "15 hp gas engine" without naming a specific brand.

It is clear, however, that stationary engines were among the first ones adapted. Horizontal stationary engines were being sold as early as 1890 by the Charter Gas Engine Company of Sterling, Illinois, on the Rock River. John Charter's son, James, left the company for Fairbanks-Morse in 1895 and a horizontal stationary engine was an early product of that company. These Fairbanks-Morse stationary engines of 12 to 40 horsepower would become one staple of the gasoline river boat. Another likely candidate, although not perhaps for the earliest installations, was the St. Marys built as early as 1903 in St. Marys, Ohio, and later in St. Charles, Missouri, by the St. Marys Machine Company. Realizing the potential of the towboat market, St. Marys even developed an ad that showed its horizontal oil engine with the legend: "The Best Engine Made for Boat Duty." The single-cylinder Bessemer, a proven performer in the oil fields, was built in Grove City, Pennsylvania, and was also used in gas boats during the first decade of the 1900s. So were the horizontal engines built in Springfield, Ohio, by Foos.

It was a situation that focused attention on stationary type engines used for marine service. In 1911, the editors at *The Gas Engine* were prompted to "question just what a 'marine engine' really is.... It seems that one type fits into certain classes of service previously presumed to belong only to the other. There is no question but what there is an increasing demand for small freight and passenger boats in many sections of the country, and the gasoline engine, stern-wheel boat is the one thing that meets this demand."

Jim Poindexter, a lifelong riverman from St. Albans, West Virginia, recalls the conversion of such engines to riverboats clearly. "The first gas boats were small towboats driven by one-cylinder gas- and oil-well engines in the 30- to 100-horsepower range," he said. "After a well was empty, some of these engines were put in a boat. They had a lot of horsepower and torque at low rpm."

The boats were functional, not grand. Said Poindexter: "Most were built right on the riverbank. Some of the more sophisticated ones may have had drawn-out plans, but most did not. Some were built from logs taken as they floated down the river. These would then go to a sawmill. Others bought their timber from the mill. Pine, oak, fir, and yellow poplar were all used. Most all these boats were stern wheelers." Although of later vintage than many gas boats of its type, the *Attaboy,* built on a riverbank in Pittsburgh in 1923, was very like the vessels that Poindexter remembered. A stern-wheeler measuring 75-feet long by 16 feet wide by 2.8 feet draft, *Attaboy* was powered by a 60-horsepower, single-cylinder Fairbanks-Morse gasoline engine.

Details of Gasoline Stern-Wheel Boats, 1911

Name	*Edith H.*	*Grace V.*	*Estelle D.*	*Gipsy*	*Sadie E.*	*Evelyn*
Model or Scow Bow	Model	Scow	Model	Scow	Model	Model
Hull Built	Unknown	1910	Unknown	1909	1899	1910
Engine Built	1907	1897	Unknown	1891	1900	1910
Engine HP	25	16	20	10	12	25
Speed	8 mph	8 mph	5-6 mph	7 mph	5½ mph	6 mph
Gross Ton	10	15	13 1/2	Unknown	10	14
Net Ton	10	12	13 1/2	Unknown	8	12
Length	65'	61'6"	62'	44'6"	50'	68'
Width	12'	14'	Unknown	8'4"	8'	12'
Wheel Diameter	9'	8'	7½'	6'	6'	7'
No. of Buckets on Wheel	12	10	10	10	12	11
Width of Buckets	14"	12"	10"	12"	12"	12"
Large Sprocket Diameter	48"	42"	39"	38-tooth	43-tooth	Unknown
Small Sprocket Diameter	9¾"	8"	11"	10-tooth	10-tooth	Unknown
Chain Size	No. 14	303	No. 120	No. 17	No. 30	Unknown
Belt Type	Rubber	Rubber	6 ply Canvas	Gandee	Unknown	Canvas

Fred Way, Jr., described the impression made upon him by such engines. They had, he recalled, "enormous firing power in those one-lung cylinders. Lying at a landing place, engine idling and paddle wheel disengaged, the sound of the exhaust was not unlike that of a huge animal 'getting its wind' after an exciting race. Huff-huff-huff-huff-BANG! BANG!... Each BANG! blew aloft two bushels of black smoke.... Such was the modest beginning of internal combustion propulsion on inland streams."

As the population, size, and sophistication of gas boats grew, so did the installation of engines expressly designed for marine applications. In June, 1913, the 138 foot-long ex-steamboat, *New Orleans,* arrived in her namesake city powered by two 50-horsepower Globe marine engines, each driving a screw propeller. The *New Orleans* was an interesting vessel. She had been built in 1911 at the Elizabeth Marine Ways Company on the Monongahela in Pittsburgh and was said to be a replica of the first Mississippi River steamboat of them all. The vessel's draft was just two feet. The Globes, built in Philadelphia by a company that had grown out of the 19th-century Pennsylvania Iron Works, had replaced the original pair of 80-horsepower steam engines. According to *MotorBoat* magazine, the "reception given to Captain Ditch's vessel was the largest ever given a motorboat in the South, if not in the entire United States."

The *New Orleans,* intended to carry freight and passengers upriver, was big for a gasoline-powered riverboat. Most, like the *St. Dennis,* built in Ravenswood, West Virginia in 1915, tended to be much smaller. The *St. Dennis* measured 60.4' long and was powered by a horizontal-engine. Such boats, as Fred Way, Jr., pointed out, "thrived and multiplied on a diet of crumbs from the steamboatmen's tables." Although they resulted in a major decline of small steamboats, the early gas boats did not pose a serious threat to larger vessels. In 1914, when the Board on Experimental Towboats of the Corps of Engineers was considering what power to use for two experimental towboats to compare the performance of a stern-wheeler and a screw-propeller

Gas boat Mary H *at Clarington, Ohio, circa 1911. Here we see the* Mary H *together with her "ferry flat" at a river landing. The* Mary H *was powered by a 15-hp engine, perhaps a Fairbanks-Morse. Fuel bills for the boat averaged 90 cents a day on nine-cent-per-gallon kerosene and $1.50 to two dollars on 12-cent gasoline. By 1911, there were over 30 boats like the* Mary H *at Parkersburg, West Virginia, alone, just a fraction of the gas boats then in operation. (Courtesy, Bud Daily)*

vessel, it settled on steam. The screw-driven vessel, the *Inspector,* was built in Charleston, West Virginia, by one the great innovators in steam, Ward Engineering.

At the time the *Inspector* was built, many steamboatmen were outspoken in their criticism of gas boats. They recognized them as a threat to jobs, and a way of life, and objected to them as explosion-prone death traps. This latter, at least, was an odd attack given the steamboat's long history of deadly wrecks and dramatic explosions known for parboiling, dismembering, and flinging victims for considerable distances. It was a fire begun on a wharf-side steamboat that spread from vessel to vessel and then leapt ashore to raze 15 city blocks in St. Louis in 1849. The attack on gas boat safety also defied statistics that indicated gasoline-propelled boats had fewer accidents than steamboats. As technology developed, however, it was not the gasoline engine in ever-larger horsepower ratings that ultimately would drive the steamboat off the rivers. It was an engine that, by 1914, was proving itself practical in ocean-going ships. It was the diesel.

By the time the United States entered World War I, inland river commerce had been on a slow decline. In 1908, president Theodore Roosevelt had appointed an Inland Waterway Commission and, in 1910, under the River and Harbor Act, the Corps of Engineers was charged to study inland river commerce and recommend improvements to equipment and channels. The War helped revitalize river traffic. As railroads simply could not move enough material, the rivers took on increased importance, beginning what came to be known in transportation circles as the "barge age." The war saw the construction of new terminal facilities, but the government took over the most lucrative routes and, after the war, growth in the tow and barge business was slow.

There were now frequent discussions about the lack of young men interested in working for the towing companies. In fact, there was little incentive for experienced rivermen to help younger people get their licenses. Even the most skilled steamboatmen earned well below what could be made ashore. And then there was the life-style itself. One writer in the early '20s characterized the crew quarters on a typical steamboat as "just large enough to squeeze into, so cold in winter that you almost freeze and so hot in summer that you don't even attempt to sleep in them."

There remained, always, the expenses of running even the most efficient steamboat. "By reason of a necessarily low piston speed the economy of paddle wheel machinery is comparatively poor," pointed out an article in *Waterways Journal* in 1924. But rivermen had come to rely on the maneuverability of steam paddle wheelers and their shallow draft.

The diesel had its problems, too. The engines were large and heavy for the power they produced, making them most suitable, beginning around 1903 in Europe, for really big, steel ships. The engines' weight and bulk was augmented for years by the necessity of carrying air compressors and air tanks needed not only for starting but for injecting the fuel. There was the also the riverman's need to be able to reverse the engine's direction without a heavy transmission, and there were issues of cost and the reliability of a "true" diesel — an engine that both started and ran without an external ignition source of any kind.

Of central importance to rivermen was the boat's drive system. Steam stern-wheel and screw-drive, and diesel stern-wheel and screw-drive were all being experimented with. Although, even before the war, it was known that the screw propeller was unmatched for sheer efficiency, paddle wheels would remain in evidence for the entire inter-war period. Diesel-boat designers also had the challenge of what method to use to turn the paddle wheel. Early worm or bevel gear reduction systems were prone to failure and usually added weight at undesirable places. Diesel-electric systems, although they would develop to be an efficient method were, initially, found wanting in terms of starting torque. Direct drive systems with transmissions were tried as were a variety of shaft- and chain-drive layouts.

By the early '20s, however, these issues were all being addressed and several American companies were producing engines of 100 to 200 horsepower and higher that were demonstrating the diesel's great promise — reliable, efficient operation at minimal fuel cost. These attributes together with the boats' greater cruising range, could no longer be ignored by towing companies, whether large or small.

"The following are cited as examples of the performances of boats of this type," reported *Waterways Journal* in September, 1924. "Recently, a 200 horse power stern wheel motor towboat towed three 200-foot and three 150-foot empty steel barges upstream and returned with a tow of similar barges loaded with 5,000 tons of cargo, covering a total distance for the round trip of 108 miles in 24 hours running time, at a total cost of $50 for the round trip, excluding investment charges.

"A compound condensing towboat of 600 to 650 horsepower with the same identical tow upstream and downstream over the same portion of the river made the trip in 23 hours at a cost of approximately $250, excluding investment charges.

"Another towboat of the same power completed a 2,300 mile trip. During this trip she consumed $300 worth of fuel, all of which was placed aboard [no need to stop for coal] before starting.

"A motor towboat properly constructed cannot be built any cheaper than a steamboat of the same power. It can, however, be operated cheaper; and therein lies the reason and justification for its building."

By 1923 or 1924 with the introduction of diesel engines that combined airless fuel injection systems known as "common-rail" systems, and reliable direct-reversing models, the diesel engine was about to come of age on America's inland river system. In a four-cycle diesel, "direct reversing" referred to the ability to reverse the engine's direction by means of sliding its camshaft so that a different, opposed set of lobes operated the pushrods and changed the valve timing. This was accomplished by stopping the engine and then using air pressure to reposition the cam. Two-cycle diesels with direct reversing were stopped but then immediately restarted by operating a lever that injected air into one of the cylinders whose piston was poised so as to reverse the crankshaft's direction of rotation when ignition occurred.

Boats so equipped were big news. Although some riverboats had been built with non-reversing engines by the early '20s, in 1923, a gravel-contracting firm in Nashville took delivery of what was called "the first reversing diesel engine, stern wheel towboat." This vessel, the *Harvey,* was 113 feet long, had a 26-foot beam and a three-foot draft. Built in Nashville by the Nashville Bridge Company, she was powered by a pair of four-cylinder, two-cycle, Worthington Diesel engines that each produced 120 horsepower at 375 rpm.

Worthington, then based in Buffalo, New York, traced its history to a steam-driven boiler-feed pump invented in 1840 by Henry Worthington in Springfield, Ohio. Although the firm was reorganized in 1916 as the Worthington Pump and Machinery Company, it had been involved in diesel engines since one of its engineers, J. F. Grace, visited the Electrical Exposition at Madison Square Garden in 1898 and saw there a 20-horsepower diesel engine. Restrained by the Diesel patents, Worthington nevertheless got exposure to the engines by manufacturing them in a Cudahy, Wisconsin, factory for the Diesel Motor Company of America, beginning in 1902.

Later, with the Diesel patents having expired, Worthington undertook its own diesel program and was credited with building the first all-American horizontal diesel in 1913. The engines in the *Harvey* were based on a Worthington series first marketed in 1920. The line was based on a cylinder size of 10¼" x 10 ½" and featured vertical cylinders, common rail fuel injection and precombustion chambers in the cylinder heads. It was a very successful innovation described this way by *Diesel Progress* magazine:

"The fuel was injected into an injection chamber called a precombustion chamber, where it was only partially burned due to the limited amount of combustion air in this chamber. Combustion was completed in the main combustion chamber as the partially burned products expanded from the precombustion chamber.... This system of combustion prevented detonation and the high cylinder pressures resulting therefrom."

Nashville Bridge addressed the drive-line challenge on the *Harvey* with a shaft that ran aft from each engine to a point some 10 feet forward of the stern. There the shaft drove a sprocket system whose chains operated a driveshaft on either side of the boat. These ran out to the split paddle wheel and drove each wheel independently via a bevel gear that meshed with another gear on the wheel. T. L. Herbert and Sons owned a steamboat of similar size to the *Harvey* and was in a good position to compare the two vessels, including their fuel consumption, which worked out to 7.9 pounds of coal per horsepower-hour versus one-half pound of fuel oil. The *Waterways Journal* recorded the performance of the *Harvey* this way:

> "The daily fuel and lubricating oil bill for the steamer is $40 per twin-hour day as compared with $9.50 for the Diesel. These figures are not estimated costs, but are from the owner's records. The Diesel boat shows additional savings in that it requires less operating force and less time for fueling. Specifically the Diesel has two less in her crew than that in the steamboat, and there is an approximate saving of 154 hours per year in fueling time. The company's records show that the steam vessel requires three hours per week as against 30 minutes fueling time every three months for the diesels. This does not include the attendant savings in labor cost, delays lost in towing time, and so forth."

It was now clear to most objective observers that the steamboat's near-century-long run was on its downside. "When a boat like the *Harvey*," said *Waterways Journal,* can be operated on less than one-half pound of oil per horsepower per hour, and when a man like her pilot, who has spent all his life on steamers, says that she is the best boat that he has had to maneuver, the possibilities for the development of the direct-drive Diesel engine installations on our rivers ought to be evident to everybody."

In May 1923, the first stern-wheel riverboats with direct-reversing *oil* engines were also launched, the *W. C. Kelley* and the *Trojan,* both powered by Fairbanks-Morse engines. The 82-foot *Trojan* was designed by the engineering department at Jones and Laughlin Steel Corporation for use around Pittsburgh. The vessel was powered by a pair of four-cylinder, 200 horsepower Fairbanks-Morse CO engines. These two-cycle engines were not true diesels. They had lower compression ratios and lower cylinder pressures than would a diesel, and early models, introduced in 1914, required a cylinder-head-mounted bulb ("hot bulb") to be heated prior to starting. Thereafter, the cylinder was kept hot by combustion. By the early '20s, the CO was started after heating each cylinder head with a built-in blow torch that was extinguished after starting.

While oil engines were not quite diesels, they did operate economically on inexpensive crude oil, although they did require a higher grade oil than a diesel. They were also less expensive than a true diesel and proved to be about as indestructible an engine as was ever built. The *Trojan* was still working in Pittsburgh in 1945. She was then owned by the Vestal Coal Company but had been converted to a single-screw boat. As for the economy of her Fairbanks-Morse CO, Jones and Laughlin estimated her expenses to be one fourth those of a comparable steam towboat.

The confusion in terminology relating to oil and diesel engines was long-lived. It was in large part the product of legal threats by Rudolf Diesel's license holders. They understandably didn't want the word "diesel" used by anyone who hadn't paid a handsome fee to the inventor. This sometimes resulted in makers of "semi-diesel" engines — machines which required an ignition source to be started — to call their machines "oil engines." As if to ensure that the confusion would continue, Fairbanks-Morse referred to the *Trojan* in subsequent ads as being a diesel.

Just as a fire in Detroit at Olds Motor Works in 1903 had given impetus to the first successful demonstration of automotive mass production, an earlier fire in Chicago had given name to what would become one of the country's most successful engine companies. The fire at Olds had been started by exploding gasoline. The fire in Chicago had been started, it was claimed for many years, by a cow. On a Sunday night in October 1871, the flames spread from the O'Leary barn — leaving the O'Leary home unscathed — leaped the Chicago River and then destroyed some 18,000 buildings.

Losses were tremendous, estimated at $200 million, and included a notable art collection and a collection of Lincoln memorabilia maintained by an old friend of the murdered president. Among the businesses lost was Fairbanks, Greenleaf & Co., the Midwest agent for a Vermont company that had invented the platform scale. When the company reopened after the fire, the Greenleaf component had left. The firm was now managed by a former apprentice in the scale company who had moved to the Midwest as a top salesman. His name was Charles Hosmer Morse. The company was now called Fairbanks, Morse & Co.

Company histories would, in later times, refer to C. H. Morse as the firm's "controlling genius." Although the precision scale business was always his first love, Morse saw big opportunity in a variety of areas. He

United Engineering Works, two-cycle, hot ball, oil engine, circa 1918. Engines like this operated on comparatively cheap fuel oil or kerosene that cost about one-fourth the price of gasoline. Together with more expensive diesels, which had a higher compression ratio and needed no ignition system at all, the "hot head" spelled the end for big, gasoline engines.

Ignition required to start the engine was provided by heating a ball-shaped projection in the combustion chamber with a blow torch attached to the cylinder head (On this United engine, the device was mounted on the starboard side.) or, later, electricity. The heat of combustion thereafter kept the ball at a temperature high enough for the engine to continue running.

United was but one of many makers of hot-ball engines that included Bolinder, Mietz & Weiss, Skandia, Kahlenberg, Venn-Severin, Dodge, and Missouri.

acquired the Eclipse Windmill Company in Beloit, Wisconsin, in 1880, and a steam engine maker in Beloit in 1893. Railroad and pumping equipment was added and later, in 1908, Morse also acquired an electric motor manufacturer. But it was for its internal combustion engines that Fairbanks-Morse would become most famous. Quick to perceive the potential of the machines, Morse acquired the rights to a horizontal gasoline stationary engine invented by a German immigrant named Franz Burger and produced in Sterling, Illinois, by the Charter Gas Engine Company.

Charles Hosmer Morse (born 1833, St. Johnsbury Center, Vermont; died 1921, Winter Park, Florida): Not many men made and kept a fortune in the engine industry, but C. H. Morse was one who did. Fairbanks-Morse engines were among the most popular used in the early generation of gasoline-powered riverboats. Those engines formed the basis for what would become a major force in the marine engine industry. (Courtesy, Coltec Industries)

Having prevailed upon Franz Burger to come to Sterling, John Charter soon decided that there was more potential in natural-gas engines than in engines that operated on gasoline. Although he'd patented the Burger engine, and a functional mixing valve, he now found himself in conflict with the company directors. John Charter left the company, probably in 1889 or '90 and, together with his son, commenced making another engine in Chicago under the name Caldwell-Charter. According to William Robinson, who grew up in Sterling, Illinois, and whose father worked at Charter and became head of the company with Charter's departure, the new venture was not a success. "They sold out their patterns and engine investments to the Fairbanks-Scale Company," Robinson remembered.

James Charter, who had continued development of the Burger-designed engine, moved from Chicago up to Beloit, circa 1893, to assist in the start of production. Those first engines were called Fairbanks-Charter for what Robinson recalled to be "a year or two," after which time the name was changed to Fairbanks-Morse

Fairbanks-Charter horizontal engine, 1893. When C. H. Morse acquired the rights to build this engine, he established the basis for a manufacturing operation whose engines would become famous throughout the world. The Fairbanks-Charter was among the first generation of gasoline engines to be designed and manufactured in the U.S. (Courtesy, Coltec Industries)

in 1895. Thus was launched the line of engines that would become known throughout the world. Development of the Charter engine by James Charter and Fairbanks-Morse engineer Franklin G. Hobart was methodical. Although Charter left in 1900 for unexplained reasons, by the early 1900s an extensive line of horizontal and vertical stationary engines was in production. It was this horizontal that was used by early gas boat builders. These engines lasted a long time. A 12-horsepower model installed in the *Elaine* in 1898 was still stroking steadily away 30 years later. Fred Way, Jr. , once chronicled a 32-horsepower Fairbanks that was installed in the *Hanover.* When that boat burned and sank, "the engine was recovered and placed on the *New Hanover,* ran ten more years, then that one burned and sank. The engine was again recovered from the bottom of the river and was put on the second *New Hanover* where it chugged along three more years."

The Fairbanks-Morse engine lineup, meanwhile, kept growing, and a Canadian branch was established in 1905. The first marine engine, the 20-horsepower, two-cylinder, four-cycle Type "B," was introduced in 1900. The line was further developed to include a 100-hp, four-cylinder version that, like other higher-horsepower Type "B" models, included a compressed-air starter. It's worth noting that the Fairbanks catalog for the Type "B" actually included installation drawings showing the drive system for a split-paddle-wheel

Fig. ME2404
40-H. P. Four Cylinder Engine—Exhaust Side.

Fairbanks-Morse Type "B" Marine Engines

COMPLETE EQUIPMENT

With each Fairbanks-Morse Heavy Duty Marine Engine we furnish one sparking dynamo attached to engine, also one electric battery, spark coil, switch, wire, reversing gear with combined thrust bearing, inboard stuffing box with gland, outboard stern bearing, stern tube, propeller wheel, propeller shaft, coupling, circulating pump with sea cock and bilge connections, gasoline and water pipes and fittings, wrenches, oilers, etc

On engines of 40 H. P. and smaller, exhaust pot, pipe and fittings, a direct driven air pump for blowing whistle, which is quite essential on large boats, and a hand starter pump for starting engine, are also furnished. No whistle or air receiver is furnished unless specially ordered.

Equipment furnished for fresh or salt water service as ordered.

H. P.	No. of Cylinders	Speed, R. P. M.	Dia. Propeller Wheel, Inches	Height from Foot Plate to Top of Engine, Inches	Height from Foot Plate to Shaft, Inches	Width of Base, Inches	Space Required Fore and Aft, Inches	Diameter of Fly Wheel, Inches	Approximate Weight of Engine, Lbs.	Approximate Shipping Weight, Lbs.	Export Shipping Weight, Pounds	Cubic Feet of Space Occupied	*Code Word for Gasoline Engine	
													Salt Water Fittings	Fresh Water Fittings
20	2	350	30	42	$3\frac{3}{4}$	25	79	32	2300	3400	4300	82	Rosy	Lantern
30	3	350	36	42	$3\frac{3}{4}$	25	95	32	3400	4500	5700	86	Ruby	Linden
40	4	350	40	42	$3\frac{3}{4}$	25	110	32	4300	5300	6500	110	Rainor	Lobby

*Code words are for Gasoline Engines. For other fuels use following prefixes before code words given: Petrol (Pe); Naphtha (Na); Distillate (Di); Kerosene (Ker); Alcohol (Al).

Fairbanks-Morse Type "B" four-cylinder marine engine, 1901. The company's first engine designed expressly for marine purposes, the Type "B" was a big seller for Fairbanks-Morse. Available in two-, three-, and four-cylinder models, the Type "B" was of the slow-speed, heavy-duty design typical of that generation engine. This four-cylinder model had make-and-break ignition, developed 40 hp, at 350 rpm and weighed 4,300 pounds. It was shipped complete with shaft tube and bearings, shaft, propeller, seacock and all plumbing fittings. (Courtesy, Coltec Industries) stern- wheeler. In addition to the Type "B," Fairbanks also developed into a full lineup of one-, two, three-, and four-cylinder, two-cycle marine engines before 1910. Company records showed that by 1915, almost 168,000 Fairbanks-Morse engines had been sold. Many of the stationary engines were bought by farmers and railroads already familiar with the company's windmills. The level of production was such that engines were produced in Beloit and, with the acquisition of the Sheffield Power Works, makers of the two-cycle Simplex engine, in Three Rivers, Michigan.

By 1910, although Fairbanks-Morse engines of all types had well established themselves in the market, there was more competition than ever before. Engineer Hobart, noting the introduction by August Mietz and C. W. Weiss of the pioneering, two-cycle Mietz & Weiss hot-bulb oil engine, recognized the commercial potential of such machines. In 1912, Hobart began development of what would become the F-M Type Y oil engine. The introduction in 1914 of a marine version of this engine, the CO gave a significant impetus to already strong sales. The CO was built in just the horsepower range then needed — 30 to 300 horsepower. It was the right engine at the right time, a time when gasoline prices were increasing and oil engines were becoming very popular. Models below 150 horsepower were equipped with reversing gears while more powerful engines were made directly reversible. "They filled such a need that they practically dominated their respective markets for years," said a Fairbanks engineer.

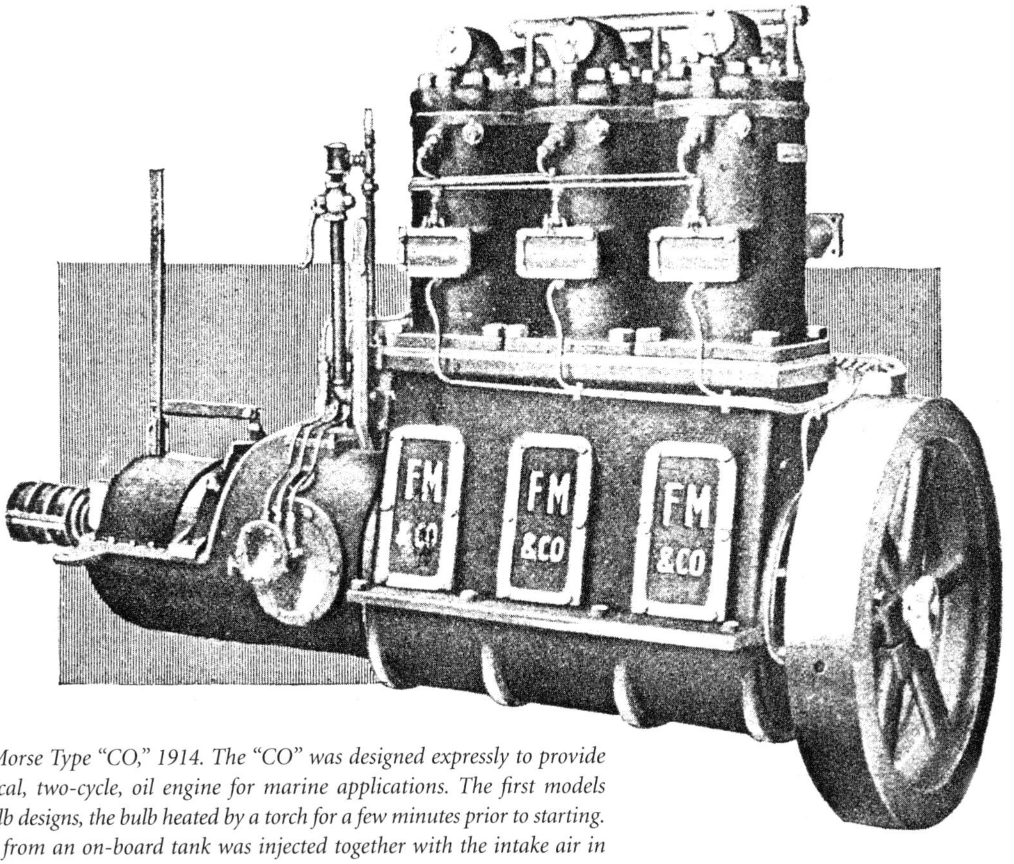

Fairbanks-Morse Type "CO," 1914. The "CO" was designed expressly to provide an economical, two-cycle, oil engine for marine applications. The first models were hot-bulb designs, the bulb heated by a torch for a few minutes prior to starting. Fresh water from an on-board tank was injected together with the intake air in order to control pre-ignition and reduce carbon buildup. Eventually, the "CO" evolved into a purely compression ignition engine — a diesel.

Fairbanks-Morse ad, circa 1912. The scope of F-M's business is suggested by this ad noting both a line of two- and four-cycle motors that were available throughout the U.S. and in England.

By now, C. H. Morse was well on his way to recognition as one of the country's most important industrialists. He built a lavish mansion in Chicago and furnished it with beautiful objects, including a variety of Tiffany pieces. Winters were spent in Winter Park, Florida, where he died in 1921. By then, the Fairbanks-Morse tradition of innovation was well established and Morse's heirs were well provided for. Years later, in 1942, his granddaughter established the Charles Hosmer Morse Museum of American Art in Winter Park, and Morse's original Tiffany glass objects are among the treasures displayed.

It was upon the CO that the first Fairbanks-Morse diesel was based. "One of the improvements consisted of the gradual raising of compression pressure with corresponding reduction of fuel consumption," recorded one Fairbanks-Morse retrospective. "This pressure by 1924 was so high that we obtained 'cold start' and we then called it a Diesel engine — our Model 32. It was followed in 1925 by a larger engine... our Model 33. Both models were developed in Beloit, Wisconsin, without the acquisition of foreign patents. They were designed specifically for the American market. With them, Fairbanks-Morse & Co. was firmly launched in the Diesel engine business."

The matter of not relying on foreign patents was a point of pride to Fairbanks-Morse as it was to the old San Francisco diesel pioneer, Atlas-Imperial. Beginning with Adolphus Busch's acquiring the American rights to Rudolf Diesel's patents and engines in 1897, it was not unusual for American companies to build under license. Besides Busch's American Diesel Engine Company, for example, there was the Hooven-

Owens-Rentchler (H.O.R.). In 1924, it acquired M.A.N patents for a big, two-cycle marine diesel. Distributed by the National Supply Company, the Superior diesels grew out of a gasoline engine company begun in Ohio, but the diesels were initially based on Deutz patents acquired in 1923 with the purchase of the Otto Engine Works in Philadelphia, Pennsylvania. When the Buda Company of Harvey, Illinois, decided to enter the diesel engine business in 1926, it did so with a high-speed engine built under license from M.A.N. The pioneering German company also licensed, in 1910, a branch of Electric Boat Company formed to build M.A.N. diesels for use in submarines. This was the New London Ship & Engine Company (Nelseco).

Fairbanks-Morse's pride as a "home-grown" diesel engine manufacturer was justified. But it might be noted that the first diesels grew out of the CO; that they did so after the expiration of Diesel's patent and that of the American Diesel Engine Company; and they were two-cycle engines rather than four-cycle type developed by Diesel. They were unique to Fairbanks-Morse, of course, and innovative. The line of marine diesels introduced during the mid-'20s included such features as oil cooling of the pistons, pressurized crankcase lubrication, direct reversing, and improved performance thanks to a separately mounted scavenging air pump. These would be features of subsequent models like the Type 37 offered in horsepower ratings up to 2,000. A smaller, four-cycle, high speed (1,200 rpm) model, the 36A was also introduced, fitted with a reduction gear. Then in the early '30s, development was begun on a thoroughly unique engine, a two-cycle with horizontally opposed pistons. It is described more fully in Chapter 5.

Fairbanks-Morse three-cylinder, Model 36A circa 1935: The 36A was FM's compact, high-speed model intended for both pleasure and commercial boats. A full line was offered, from one to six cylinders, but all were based on the same 4¼" bore dimension. A Joe's gear was fitted for reversing. (Courtesy, Coltec Industries)

The appearance on the rivers of boats like the *Harvey,* the *Trojan,* and the *W. C. Kelley,* and their well-documented performances, foretold the future of the riverboat. In Charleston, West Virginia, at the Ward shipyard, which had achieved widespread fame for improvements made to boilers and for its other steam-related advances in the 1880s, diesel-powered boats began slipping down the ways into the Kanawha River in the mid-'20s. At one time, Ward had offered 120 models and sizes of the water-tube boilers it had pioneered. Although Ward continued to build steam turbine-powered boats, it now began an increasing use of diesels manufactured by Fairbanks-Morse, Cooper-Bessemer, Atlas-Imperial, Winton, and others. The Winton-powered 119-foot *J. B. Battle,* with its diesel-electric drive, pointed to a system that provided an easy method of reversing the engine without stopping it momentarily to operate a direct-reversing mechanism.

Finally, in 1931, after some two decades of pros and cons relating to steam and diesel, stern wheel and screw propeller, the future arrived, unalterably and dramatically. That summer, in Iowa, the Dubuque Boat & Boiler Co. launched a vessel that surpassed the biggest steamboat to become the world's largest river towboat. This was the tunnel stern, 226-foot *Herbert Hoover.* She was driven by a pair of eight-foot diameter screw propellers powered by two 1,100 horsepower diesels built by McIntosh & Seymour, a division of the American Locomotive Company. These giants were four-cycle engines with 20-inch cylinder bores and a 24-inch piston stroke. They were rated at 200 rpm.

In March 1945, riverboat enthusiast Frederick Way, Jr., published his first *Inland River Record,* a compilation of vessels then in operation on the Mississippi River system. In the *Record,* Way listed as many boats as he could document, including such basic data as name, dimensions, year and place of construction, wood or steel, stern wheel or screw propeller, and the type of engine. When he was finished, Way added up the total horsepower of steam- and diesel-powered boats and found the former still led by some 25,000 horsepower.

"The author is first to gracefully concede," wrote Way in his introduction, "that the war situation is responsible for the preponderance of steam horsepower. Well known is the fact that the 21 towboats of the Defense Plant Corporation, each rated 2,000 h.p., were originally designed as Diesel craft. Thus we have 42,000 horsepower added in the past two years to the steam horsepower, due solely to the fact that Diesel equipment was not available. Were it not for this one circumstance, the Diesel horsepower would today have a strong lead."

Way's publication proved to be of interest not only to the rivermen of that time, but it would also serve as a historical record of the progression of internal combustion on the rivers. Consider the 77-foot sternwheeler *Abigale,* built at Shawneetown, Illinois, in 1924 as a steam towboat. She was converted sometime thereafter in Kansas City, her steam machinery replaced by a Caterpillar diesel. By the early '20s, not many builders of smaller towboats were even considering steam power. When the 55.6-foot *Dorothy T* was launched at Paducah, Kentucky, in 1923, a 70-horsepower Wolverine diesel, built in Connecticut by a company that owed its founding to internal combustion pioneer Clark Sintz was mounted in her wooden hull. The *Record* also shows that, after 1920, the oil engine and the diesel had largely displaced the gasoline engine, although smaller towboats with gasoline engines were being built throughout the interwars period.

A wide variety of oil and diesel engines was used. Although Fairbanks-Morse always had a large portion of the towboat market, first with oil engines and then with its diesels, most engine manufacturers achieved at least some share of the towboat operator's business. Atlas-Imperial diesels were used in bigger towboats,

Fairbanks-Morse, six-cylinder, Model 37. With a 12" bore and 15" stroke, this engine was popular for towboats and tugs. The big, cylindrical housing at the engine's front housed a centrifugal pump used to scavenge (provide fresh air) to the cylinders of this two-cycle diesel. Said the company: "F-M simplicity means no inlet and exhaust valves in the cylinder head. No periodic valve grinding. The fuel injection system is simple and trouble-free — a mechanical injection system." The Model 37 was direct-reversing, the process controlled by the handwheels on the engine's port side. (Courtesy, Coltec Industries)

as were two-cycle Kahlenberg semi-diesels made in Two Rivers, Wisconsin. Wintons, Superiors, Worthingtons, and Nelsecos were used in some numbers and so were Bethlehems. Cummins and Buda diesels found a market in smaller boats. Among the latter was the *W. L. Heckmann,* a 44-foot, screw-propeller boat named for Captain William L. Heckmann, who was known to the readership of the *Waterways Journal* as "Steamboat Bill." Big diesels from Busch-Sulzer in St. Louis were sometimes specified for bigger vessels, like the twin-prop, 154-foot *W. W. Marting.*

Few of these manufacturers had any connection to the steam engines that they had so thoroughly displaced. One that did owed its existence to the merger of two distinct companies. One had been founded in 1833 as an iron foundry in Mt. Vernon, Ohio, by Charles and Elias Cooper — C. & G. Cooper Co. The other was

Atlas diesel. It was a classic, slow-speed Atlas like this one for which Cooper-Bessemer acquired the rights and began selling in 1923. (Courtesy, Dan Webster)

begun by in 1899 by a mechanically oriented physician named E. J. Fithian and a machinist named John Carruthers in Grove City, Pennsylvania, some 60 miles south of where Edwin Drake's first oil well had been drilled in Titusville in 1859. This was the Bessemer Gas Engine Company.

The merger of these two successful firms in April 1929, illustrated the challenge facing any company wanting to establish itself in the diesel engine field. With rare exceptions, any venture into diesels required significant capital for research and development, experienced engineering talent, and an extensive manufacturing facility. By 1929, both Cooper and Bessemer had progressed about as far as was practical as individual entities. Cooper, in particular, faced a major hurdle. The company had built locomotives during the Civil War and not long thereafter acquired the license to the steam engines developed in Providence, Rhode Island, by George H. Corliss. The Cooper-Corliss was, as the company liked to say, "the first Corliss Engine built in the West." Some of these machines were 200-ton giants, with pistons seven feet in diameter and flywheels that weighed more than one of Cooper's locomotives. The company also began manufacture of portable steam engines for use on farms and in 1875 introduced a highly successful steam traction engine.

None of this would be enough, the company recognized, for it to remain viable long into the 1900s. Development of steam turbines by Westinghouse and the apparently limitless potential of gasoline and natural-gas engines meant that Cooper would need to head in a new direction. The first step was taken with the introduction in 1909 of a huge, natural-gas engine and compressor for use on a pipeline in West Virginia. The manufacture and installation of such machines after World War I, together with smaller 80- to 170-horsepower models, made Cooper the leader in pipeline compression engines. Its field engineers formed a rugged and romantic group known as "erection men" who somehow managed to get the largest of engines put together and operating in the remotest of places. To advance its diesel prospects, meanwhile, Cooper, in 1927, acquired the rights from Maschinenbrik Augsburg Nurnburg (M.A.N.) to that company's four-cycle diesel engine and began investigating its potential development as the Cooper-M.A.N. But the company still felt vulnerable. "Cooper did not dominate small engine production for gasoline extraction," wrote David N. Keller in his *Cooper Industries 1833–1983.* "That distinction went to a competitor. . . ."

In fact, while Cooper was busy building engines for the natural gas industry, to the East, in Pennsylvania, Bessemer was already well-known as the builder not only of engines for use in oil fields, but for marine diesels as well. The company had come to dominate a good portion of the market for gasoline engines but Fithian, observing the impact that oil engines and then diesels were having, concluded that Bessemer's future lay with the diesel. His associate, John Carruthers, the man who made the factory work and who had developed the Bessemer engines, died in the early '20s and Fithian then purchased Carruthers's share in the company from his two sons. He promptly acquired a license from Atlas-Imperial in Oakland, California, and the company began development of what would be introduced in 1923 as the Bessemer Atlas Diesel. These were similar to the California engines but carried the legend, "Bessemer Atlas Diesel" on the inspection covers. Three-, four-, and six-cylinder models were available in ratings of 50 to 1,000 horsepower.

"For thirty years they have been Pacific Coast favorites," said a company ad in December 1923, "and are now brought to the East Coast, the Great Lakes, Gulf and Eastern rivers through the huge plant of the Bessemer Gas Engine Co., one of the largest manufacturers of gas and oil engines in the world."

The first customer for one of these engines may well have been the Borough of Grove City itself, which purchased a stationary, 360-horsepower model on November 21, 1923. Only four engines were sold in 1923, but the following year, more than two dozen were built. A number of these were bought for use in fishing schooners in Boston and Gloucester and another, a 230-horsepower, six-cylinder was installed in the yacht *Speejacks II,* successor to the *Speejacks* which, in 1923, had become the first gasoline-powered yacht to circumnavigate the world. In 1925, Packard president Henry B. Joy ordered a Bessemer for his new yacht, *Spray III.*

While it found ready customers among wealthy yachtsmen and New England schooner owners, Bessemer did not make inroads into the river towboat fleet until 1926. That June, the Iron City Sand & Gravel Company in Pittsburgh ordered a three-cylinder, 60-horsepower Bessemer for its new *Poppy* and a similar engine went into another towboat, the *Virginia.* If the towboat was to be a stern-wheeler rather than screw-driven, a stationary version of the Bessemer was ordered. For example, in January 1927, Charles Zubik, among the most colorful and independent of Pittsburgh's towboat operators, purchased a 110-horsepower, three-cylinder Bessemer for his 71-foot *Tornado* and a 235-horsepower, four-cylinder for his 73-foot *Smoky City.*

Wonderful as it was, the Atlas-Imperial was recognized by Bessemer engineers as representing World War I-era technology. After the war, the great cast iron marine gasoline engines with individual cylinders and exposed pushrods quickly gave way to ever more streamlined models as marine engine companies looked to automakers as role-models for such things. In December, 1927, Bessemer introduced its R-line of engines with en bloc cylinders "strongly reminiscent of automotive practice."

The six- or eight-cylinder R was a world-removed from the Atlas-Imperial, and it was of fabulous specification by any measure. It had what Bessemer called an overhead camshaft, although the cam was actually mounted alongside the top of the cylinder head rather than directly above the valves, replaceable cylinder liners, and an enclosed flywheel. Mounted in removable cages in the cylinder heads was a pair of forged-steel exhaust valves and a pair of inlet valves — four valves per cylinder! The 10-inch diameter crankshaft was built to Lloyd's specifications. It was unusual in that it was forged in two sections, which were bolted together. Each crank section served one block of cylinders; the crankshaft itself was carried in main bearings 12 3/4 inches in diameter. A pinion mounted at the coupling drove the camshafts through a 2:1 reduction gear. This feature contributed to the engine's comparatively compact dimensions and to its smooth operation. When running at 300 rpm, the engine was said to be as "quiet and relatively free from vibration as the most delicately balanced 8 cylinder automobile engine."

The six- and eight-cylinder R series engines shared 14" x 18" bore-stroke dimensions, the six producing up to 675 horsepower, the eight developing up to 900 horsepower, both at a maximum 375 rpm. They were, of course, direct-reversing, and they had a motor-driven lubrication system that included an oil cooler. An engine-driven fuel pump delivered fuel under pressure through a common rail to the injector valves at each cylinder. These, as was typical of common rail systems, were operated by the camshaft. These R models found their primary market in commercial vessels and big (170 to 180 the foot) commuter yachts designed by Henry

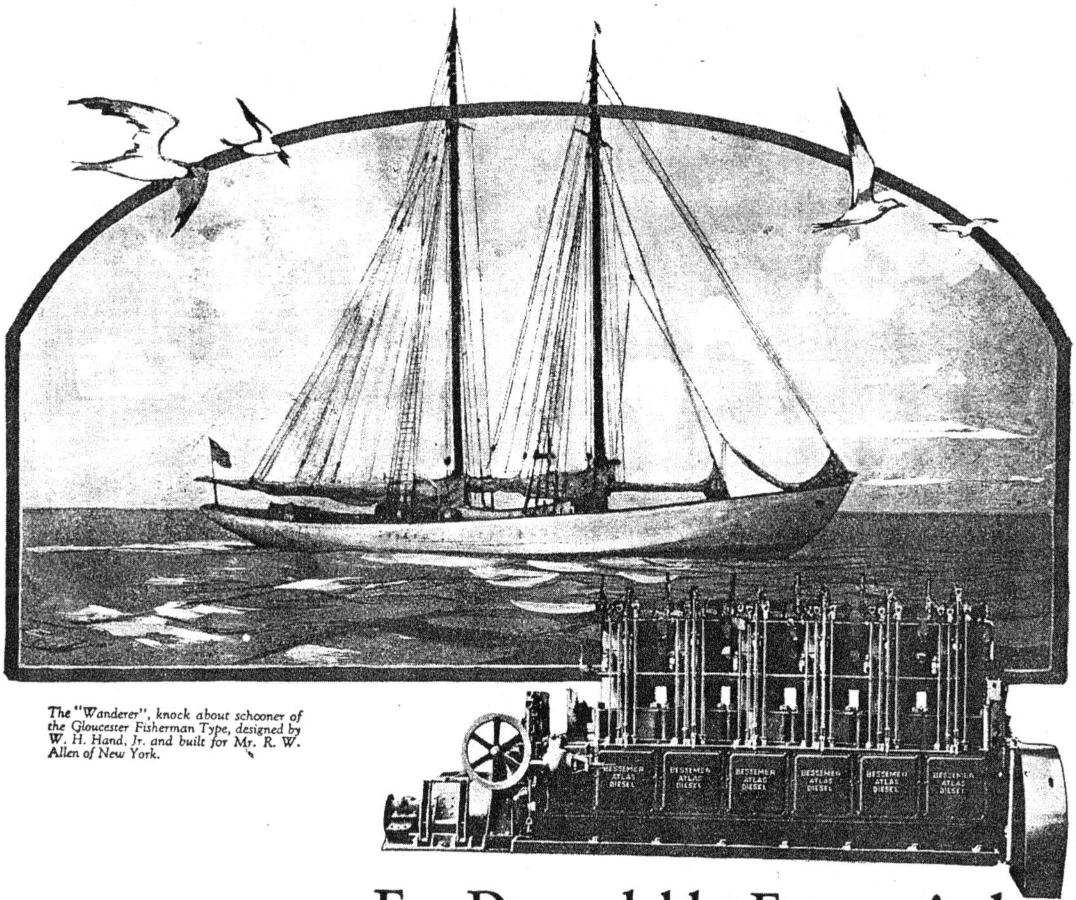

The "Wanderer", knock about schooner of the Gloucester Fisherman Type, designed by W. H. Hand, Jr. and built for Mr. R. W. Allen of New York.

For Dependable, Economical Marine Power Plants

Bessemer Atlas Diesel Engines are of the four cycle type built in sizes from 50 H. P. to 1000 H. P. in three, four and six cylinder styles—there is a model for every marine requirement.

Fuel injection is of the "airless injection" type, by pressure. This eliminates the costly and dangerous high pressure air compressor and permits the highest economy in fuel consumption. They are the most economical Diesel engines built today from point of view of first cost and operation cost.

BESSEMER ATLAS DIESEL ENGINES are engines of proven superiority. Simple in design, sturdy in construction and dependable and economical in operation, they are the ideal power plants for marine use where reliability and economy are vital factors. Their efficiency is clearly indicated by their choice by many well-known designers, a recent example of which is the selection of Bessemer Atlas Diesel for the new schooner "Wanderer" designed for R. W. Allen by W. H. Hand, Jr.

For thirty years they have been Pacific Coast favorites and are now brought to the East Coast, the Great Lakes, Gulf and Eastern rivers through the huge plant of the Bessemer Gas Engine Co., one of the largest manufacturers of gas and oil engines in the world.

Write for complete information which will be gladly sent on request

THE BESSEMER GAS ENGINE COMPANY
10 Lincoln Avenue ∴ ∴ Grove City, Pa.

BESSEMER

ATL-AS DIESEL

J. Gielow in New York. The Atlas-based Bessemer, which continued to be manufactured, and which must have been considerably less expensive, was the model ordered by towboat and fishing vessel owners.

Bessemer recognized that its magnificent R was not destined for volume production. Although the R may well have been the basis for the company's reputation as builder of "the Cadillac of diesel engines," Bessemer introduced a line of smaller displacement models in 1929. The EP-6 and EP-8 shared 7½" x 10"bore/stroke dimensions and developed their 150-200 horsepower and 200-275 horsepower ratings at a comparatively high 500-670 rpm. Three- and four-cylinder versions were also available. A larger six- and eight-cylinder version, the FP, had a nine-inch bore and 12-inch stroke. The company now spoke of having improved power-to-weight ratios to 50:1 or 70:1, a progressive move. These engines were different in their architecture from the R. Most notably, the camshaft was now located in the block, operating the overhead valves — two per cylinder — by lifters and pushrods. The EP was fully enclosed and its one-piece block, like the R models, carried a replaceable liner for each cylinder. The crankshaft, still built to Lloyd's specs, was now a one-piece forging. Full-pressure lubrication remained. In December of 1929, Bessemer announced that 41 of its diesels were scheduled for installation in vessels for the upcoming season. In Grove City, the factory now covered 12 acres and contained 734 machine tools, two dozen overhead cranes, and it employed 1,400 people.

Cooper-Bessemer, Grove City, Pennsylvania, January 1938. The crankshaft for an eight-cylinder engine is installed on a double-ended lathe and all eight crank pins are being lapped at the same time. (Courtesy, Knox County Historical Society)

In the early '30s, Bessemer, under the Cooper-Bessemer name, introduced another new series, the N. Now, the cam was moved higher in the block, operating a short pushrod via a roller follower. A patented Cooper-Bessemer injection pump contained twin plungers, each replaceable and each capable of supplying the camshaft-driven injectors with adequate fuel pressure. As on the P models, single intake and exhaust valves were used, and the cylinder heads were individually removable for servicing, weighing 435 pounds apiece.

In 1930, a larger displacement version of the N, the J model, was introduced. Offered in six- and eight-cylinder models, the J had all its iron parts cast of Meehanite, although specific engines that would run continuously at higher rpm had alloy pistons to save weight. The JS had 13" x 16" bore/stroke dimensions; the JTB 12" x 15". These engines and others, including the larger displacement L models, were steadily developed by Cooper-Bessemer, each one carefully built to each customer's specific application needs.

Always conscious of using the highest quality metal alloys in its engines, the company now replaced its previous nickel-steel cylinder liners with liners made of proprietary Meehanite, an exceptionally tough, close-grained metal. Various displacements were offered, the GN based on a 10½" x 13½" bore/stroke dimension, the EN 8" x 10½". Later, a smaller version, the BN was also offered.

Cooper-Bessemer GN-6DR with sailing clutch, circa 1936. Among the features of the engine, Cooper pointed out, was that "the power stresses are carried by high-strength steel tie bolts, which extend from the tops of the cylinder heads through to the bottom of the main-bearing anchorages." This engine, with a 10½" bore and 13½" stroke was available as a six- or eight-cylinder and was a mainstay of the company's production. (Courtesy, Knox County Historical Society)

Cooper-Bessemer, Type GS-6-DR with sailing clutch, 1944. An evolution of the mid-'30s Type GN, the GS retained the same 10½" bore and 13½" stroke and a hefty crankshaft that was 7½" in diameter and exceeded Lloyd's specifications. But the GS had four-valve cylinder heads and a variety of other refinements. The instrument panel included a tachometer and gauges for fuel-oil pressure, lubricating oil pressure, and air pressure. For direct-drive installations in yachts, the engine was rated at 750 rpm. For commercial service, a speed closer to 500 rpm was generally specified to improve engine life. The six-cylinder GS developed 550 hp at 750 rpm. This particular engine was destined not for a tow boat but for the Boston fishing schooner Gertrude Parker. Its price was $19,724. (Courtesy, Knox County Historical Society)

A company memo of January 1932 gives a brief glimpse into the development of the little — 5¾" x 7¾" — BN model and reveals that, even with considerable experience behind them, Cooper-Bessemer engineers could take nothing for granted. Said the memo: "It was anticipated that there would be a considerable amount of experimental work on fuel injection, but it was not expected that there should be so much trouble encountered. The engine was first started on July 1, 1931. Considerable difficulty was experienced with fuel timing and a day was lost in getting the distributor set for the desired point of injection. In the meantime a crack developed in the cylinder block and considerable time was lost in making patches." A cross-rib was added to the block casting; new injector nozzles with six rather than five holes were installed and the injector angle changed, and a steel crown with a revised shape was used atop the aluminum pistons.

In addition, by 1936, the company had developed a pressure relief valve that, it claimed, eliminated that persistent bane of common rail injection, the "dribble" of fuel that continued even after the injection process was supposed to have ended. This caused carbon buildup, rough running, and eventually plugged the tiny holes in the injector nozzles. According to those whose engines had this improved system, it worked very well, delivering smooth operation throughout the engine's operating range.

The quality of the Cooper-Bessemer engines was reflected in their price. In the mid-'30s, a four-cylinder EP cost some $8,000 at a time when wages were often some 65 cents an hour. A few years later, an eight-cylinder JS sold for nearly $27,000 in 1942 and nearly $32,000 in 1946. But the engines remained in production throughout the harsh years following the stock market crash, and Cooper-Bessemer sales even increased to nearly a million dollars in 1930 before dropping to under $300,000 in 1931 and $66,000 in 1932. After that, things slowly began to improve. In his *Inland River Record*

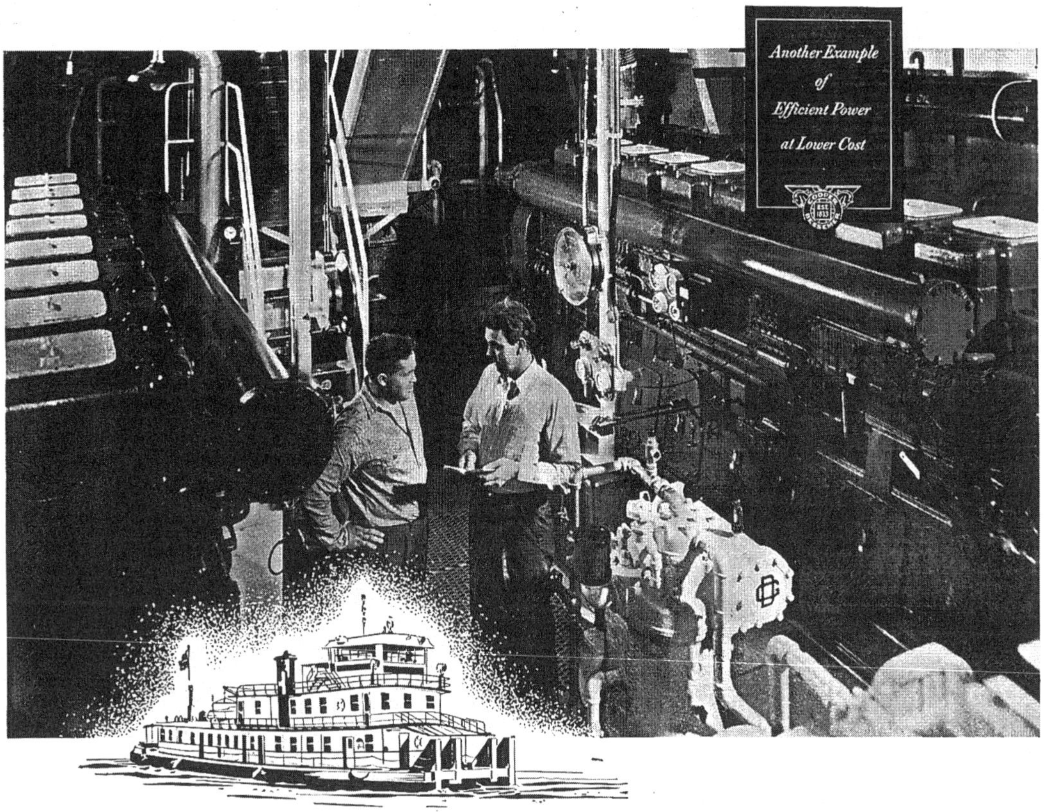

Maybe this river boat engine room holds the answer to your power problem

IF you have a stake in marine, locomotive, or stationary power, the engines shown here could well be your cue to important advantages and savings. Supercharged Cooper-Bessemer diesels, they power a big Mississippi towboat, the Frances M. Hougland. They enabled this workboat to complete 57,170 miles of tough push-towing at exceptional speeds, delivering 1,400,300 barrels of gasoline, before her first overhaul! That's an outstanding record in gruelling river service.

Cooper-Bessemer oil-burning diesels, gas-diesels and gas engines are continually setting performance records in *all* kinds of service—not only new endurance records, but record-breaking lows in fuel consumption and over-all operating cost.

It is the direct result of the intense development work that goes on all the time at Cooper-Bessemer, resulting time and again in important contributions to low-cost power. For instance, are you familiar with the Turboflow gas engine and its previously unheard of fuel economy . . . or the latest gas-diesel refinement that assures unequaled efficiency even at fractional loads?

If you want to know how you can cut your power costs year in, year out, find out about the *new* things being done by one of America's *oldest* engine builders.

The
Cooper-Bessemer
Corporation
MOUNT VERNON, OHIO GROVE CITY, PA.

DIESELS • GAS ENGINES • ENGINE-DRIVEN AND MOTOR-DRIVEN COMPRESSORS • HIGH PRESSURE LIQUID PUMPS

Proof from The Griswold-Eshleman Co., Cleveland

compiled in 1945, Fred Way, Jr., listed a total of 31 towboats powered by Bessemer or Cooper-Besemer engines, about eight percent of the total. The majority of these boats were built between 1939 and 1943. Production began to dwindle during the '50s, however, and the last marine engine was built, as nearly as anyone can now remember, in the early 1960s.

Cooper-Bessemer erecting department, Mount Vernon, Ohio, July 1944.

Like a Maine lobsterman or a Chesapeake Bay crab or oyster fisherman, most rivermen were born to their trade. Charles L. Daily was literally born on the river, in 1894, on the Ohio where it makes a number of big, sweeping bends near Vanceburg, Kentucky. "He was born on the family houseboat while it was being pooshed by the gas boat," said his son, Bud Daily. "He never had a birth certificate and he never slept ashore 'til he was 21 years old. On the river, there was nothing he couldn't do. He was the best deckhand, the best engineer, the best pilot. He could figure out any problem."

Charles L. Daily's father was, of course, also a riverman. He was a steamboatman before he built an early gas boat, using a converted International Harvester horizontal engine in 1902. This boat he used to push a flatboat full of dry goods from one town to another, selling his wares. Later, he ran a steam ferry at Steubenville, Ohio, quitting finally after 14 years without, his grandson recalls, a single day off. Then he built a gasoline pushboat and hauled barges for Dravo in the days before the company had its own fleet. And *his* father had been a coalboat pilot, guiding a barge from Pittsburgh down the rivers to New Orleans. There, the coal was sold and he would return to Pittsburgh to begin the process once again.

So the river was in Charles Daily's blood. He was the kind of riverman who could take a tow down a tricky stretch of river on a dark night always allowing for current and depth and water level and wind and whether his tow was loaded or unloaded. He always knew, where he was and what to do. If things did turn sour, he knew where and when to put his boat aground and how to fix any damage. He could line up machinery in a floating boat using the only tools that would work, a string-line and a square. He knew what size engine would be required for a new boat and how big a stern wheel would be required, the latter a skill that always depended on practical knowledge. He knew it was impossible to keep equal tension on the two drive chains going to a stern wheel. He understood that you didn't try to adjust a boat's hog chains unless you were putting the craft onto the ways or there was an accident. He was a master of the art of caulking — oakum first, then cotton, then oakum.

"He could recognize a good 'corker' by the sound of him hitting that chisel,' said his son, Bud, who grew up helping his father on pooshboats.

He could assemble a perfect barge or a perfect boat right on the riverbank with hand tools. Once in a narrow channel with barges, he came too close to something and tore the rudders off, Without rudders, he alternately loosened and tightened the face lines binding the pushboat to the barge, and backed and loosened the other and backed. Said Bud Daily, "He did this until he worked that thing in against the bank and got a line on a tree ashore. Then he fixed the boat. I mean, he just *knew.* That's all. He was, without a doubt the best there ever was."

In 1941, Charles Daily looked at a trim, 62-foot towboat called the *Edna* with the notion of buying her. Instead, the boat was bought by the River Sand Company of Steubenville, but Daily was frequently put in charge of running her. The *Edna* was a boat known on the river for having more poosh than her size would indicate, and she was a favorite of all those who ran her. During the summer of 1941 and from the spring of '42 until he entered the service, Bud Daily worked for his father aboard the *Edna.* As always on a towboat, there were many jobs to do, but Bud Daily's favorite was seeing to the boat's engine, mounted sideways up forward with long drive belts running back to the jackshaft. The engine was a four-cylinder Cooper-Bessemer and Bud took great pleasure in keeping the machine spotless.

When asked for his impression of that diesel, Bud didn't even hesitate. "Cooper-Bessemer," he said, "the Cadillac of diesel engines."

Knee-deep in herring, Havre de Grace, Maryland, May 1943. (Courtesy, Mariners' Museum, Newport News, Virginia)

Chapter 3

Fishing

"Boston is one of the best markets for gasolene motors in the United States. . . . Fishermen have arrived at an understanding with the marine motor, the result being a warm friendship destined to last."

—*MotorBoat,* April 10, 1910

"These engines were first used in the Great Lakes Districts in 1898 and, until 1904, were a product sold almost exclusively to fishermen."

—from the catalog of Kahlenberg Bros. Co., 1916

"Seattle, which is by far the greatest fishing port in the United States, so far as the value of fisheries products landed is concerned, is also the largest market in the country for gas engines, almost every make of motor in the country being represented there and the annual engine sales totaling over a million dollars."

—*Pacific Motor Boat,* May 1922

In the fall of 1897, a fisherman in Two Rivers, Wisconsin, decided that he would build a new 35-foot Mackinaw-style boat and equip it with a gasoline engine. The fisherman's name was John LaFond. Those who recorded LaFond's decision to take this revolutionary step never detailed precisely how it was made. But they did note that LaFond had, just before the event, been prospecting for gold in the Klondike. Perhaps it was there, on the Yukon River, that he saw one of the Sintz-powered boats built by Ora Mulford's Michigan Yacht and Launch Company. What is certain is that John LaFond would always be remembered in Two Rivers as the man who demonstrated to his colleagues the potential of internal combustion engines in fishing boats *and* as the grandfather of a United States senator.

Back East, at the same time that John LaFond's boat was being built, other fishermen had become aware of the gasoline engine and what it might represent to them. In Gloucester, where the great dory schooners packed the wharves of the harbor, the fleet's top mackerel fisherman, "Killer Jacobs," decided that an engine could make his *Helen Miller Gould* an even more deadly vessel. In 1898, the schooner was fitted with a four-cylinder, 25-horsepower Globe engine and, for the first time, a Gloucester schooner moved out past Eastern Point for the Grand Banks with a gasoline engine driving her. This experiment was so successful that a big, six-cylinder, 125-horsepower Globe was fitted for the 1899 season, but the schooner burned and sank at Sydney harbor in Cape Breton before they ever really got to try her out. It didn't matter. Globe engines, built at the Pennsylvania Iron Works in Philadelphia, began to be installed in other Gloucester vessels as did Standards built at the Standard Motor Works in New Jersey. Even Sterling, the Buffalo, New York, maker best known for its high-performance racing and runabout engines, was represented. The company's less glamorous, heavy-duty models found a home in the engine room of Gloucester schooners like the *Elva M. Spurling.* This Burgess-designed 80-footer was built in Essex, Massachusetts at the prolific Storey yard and was equipped with a 60-hp Sterling Model D that could drive the vessel at 7½ mph.

The new machines were run and maintained by fishermen "who saw the immense future of the internal combustion engine and prepared themselves for it," reported *The Gas Engine* in February 1914. "They have remarkably little trouble with their engines and keep them in fine running condition all the time." Besides the engine's obvious benefit in a calm or in light winds, it often enough proved to be a decisive factor in getting the catch, still fresh, to the markets in Boston and New York. In the early winter of 1911, for example, the schooner *Motor* was dismasted 150 miles offshore but "chugged merrily into Boston with a full fare of fish and from there to Gloucester for a new set of spars." A couple of years earlier, the *Bessie M. Dugan* became famous for catching $3,500 worth of pollock in just eight days of fishing. The economics of the powered schooner were irresistible. A calm would leave sailing vessels motionless while their motor-powered counterparts raced to market with their catch. An auxiliary schooner reaching Boston's T-Wharf could easily command an additional three cents per pound more than the sailing schooners would get when they finally reached port. In his article, "Boston's Commercial Motorboats," published in *MotorBoat* in 1910, George Story Hudson noted that "one calm spell made it possible for several auxiliaries to earn enough almost to pay for their motors, while the crews of the lucky craft averaged about $40 for four or five days' work with trawls."

Once begun, the move to gasoline in the schooner fleet was pretty much unstoppable. In his 1909 report on the local fisheries, the secretary of the Boston Fish Bureau noted that fishermen "must install motors or find themselves falling behind in the competition for quick trips and remunerative prices." George Story Hudson was reporting in 1910 that "Hardly a week passes without one, or perhaps, two... schooners being equipped with motors of varying power and cost. Some of the little fellows get along comfortably with 20-horse while the larger vessels find 40 up to 80 horsepower none too large to meet the requirements." Just nine years later, when the U.S. Commissioner of Fisheries issued its report for 1919 on the New England states, it found there were only 54 pure sailing schooners left, while over 300 vessels carried auxiliary power or were classified entirely as motor vessels.

According to the best recollections, the first draggers in New England waters appeared around 1895 in Provincetown, Massachusetts. These boats, two or three sloops, it appears, went after flounder, towing with the wind and tide behind them. It was a modest beginning but the market for flounder proved to be excellent and more vessels were fitted out for dragging. Then, about 1905, the fishermen began installing 10-horsepower Lathrops in their boats and, according to the *Atlantic Fisherman,* "This increased the earning power of the boats and made the business more profitable because they could cover more ground in a day by running back over their tows against the tide quicker. The adoption of power also allowed the first improvement in the method of handling the gear."

Now, from Connecticut to Maine, fishermen began installing engines and dragging under power. It was demanding work, typically carried out by boats 35-50 feet long. Within just a few years, a variety of engine builders had gained favor. In Maine, around 1910, the preference was to use a pair of two-cycle engines, one being equipped to drive a capstan as well as a propeller shaft. Sometimes, fishermen installed three engines, which suggests that the fishery represented true "gold-from-salt-water" as the phrase then went. One fisherman bought a trio of 6 1/2-horsepower Mianus motors for a twin-screw boat and kept the third motor aboard as a spare.

The *Polly,* built in Camden, Maine by H. L. Maker was considered, in 1910, to be state of the art for a small scallop dragger. She carried a five-horsepower and 6 ½-horsepower Knox, two-cycle engines, also built

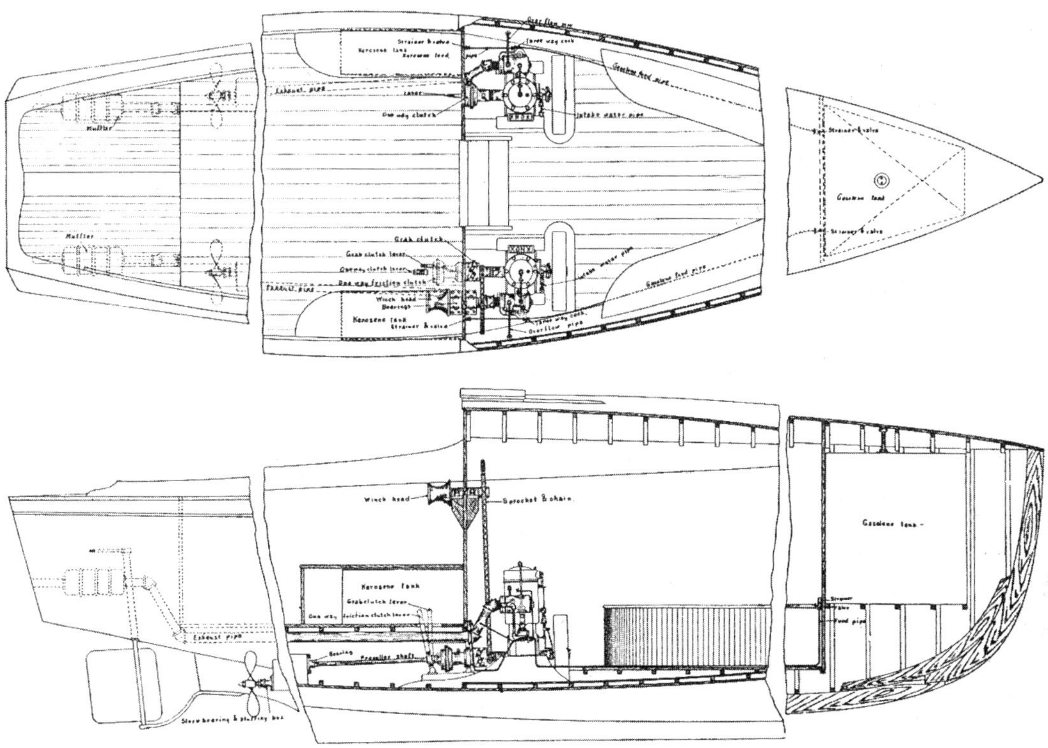

The scallop-dragger, Polly, *1910. Built at Camden, Maine, by H. L. Maker,* Polly *was powered by a pair of Knox two-cycle engines. To port was a 5-hp Knox used only for propulsion while the 6 ½ hp starboard motor was used both for propulsion and to drive the winch head used for hoisting the drag.* Polly's *50-gallon-per-day catches couldn't be matched by sail-driven draggers, and this flush-deck boat was the pride of the Camden fleet.*

in Camden, and these two motors led a hard life, running at peak power for 10 to 12 hours a day, pulling her 400-500 pound dredges. But a three-man crew on a boat like this might well land 50 gallons of scallops a day (at a market price of $.75 to $1.25 per gallon in 1910), something that would have been largely impossible with a sail-driven dragger.

If the work was hard, it was not without some natural compensation. In the fall of 1910, reporter John Bird described a day aboard *Polly* in *Motor Boating:* "It is not a bad outing at all for those who enjoy the water to make a day's trip to the scallop grounds, and it is frequently done in good weather. The boys like company and there is plenty of room in the boats for an extra man. The cabin is ample and comfortable, being supplied with a small coal stove for cooking and heating purposes. The regular crew are usually willing for the novice to help shuck the scallops and handle gear, and he is sure to get an appetite with which to even up when dinner time comes, and he is called below to a layout of fried scallops that haven't been off the bottom for more than thirty minutes. Verily, the life has its charms."

In 1904, it was estimated that 16 percent of the 9,476 fishing boats in Maine, New Hampshire, and Massachusetts were equipped with engines. Down in Connecticut, the 17.5 percent increase in the lobster catch between 1902 and 1905 was attributed to the coming of power. In Maine's Casco Bay, the elegant-

Hampton boats at Portland, Maine, fish pier, 1925. In the winter of 1900, one of the first marine engine dealers in Portland is reported to have watched the Hamptons rounding the breakwater under the impetus of their long ash sweeps and predicted that "Inside of five years, I am going to make a miniature steamboat out of every one of those poor fellows: I am going to revolutionize their lives." At that time, a fisherman might have left his mooring at one or two a.m. in order to reach the fishing grounds. After a long day and five or six hours sleep, he was back in his boat. The coming of power, first one-lungers of 6 or 8 hp, eased the back-breaking routine. The boats grew in size from 16 to 20 feet to 28 or 34 feet and, by the mid-'20s, a 38-footer was built. The larger boats were generally powered by a two-cylinder, 20-hp, four-cycle engine.

looking and able Hampton boats that had so long fished under a sprit rig, proved to be a natural for conversion to power, because their sharp bow, flat floor and rather broad stern already had similarities to the motor launch of that era. Power probably came to the black-painted Hamptons as early as 1902 or '03 and, as was always the case with the early fisherman, one brand predominated. In this case, it was the Connecticut-built Mianus that, in 1902, was available in two-, four-, and six-horsepower two-cycle models and a 12- and 24-horsepower, four- cycle. The line expanded rapidly.

By 1905 or '06, a husky, single-cylinder, nine- or 10-horsepower Mianus equipped with make-and-break ignition proved ideal for a 22- 25-foot Hampton, and could drive such a boat at about seven or eight knots. Placed well aft in the boat, the Mianus was exposed to spray and to rain, yet performed well — the perfect environment for a classic make-and-break one-lunger. In 1906, *MotorBoat* reported on a Hampton fisherman who replaced his Mianus with a two-cylinder, four-cycle engine that was soon "coated with thick oil and fish scales, and the bright parts rusted badly by salt water." When the engine failed him and he needed a tow from a steam-driven fishing boat, the fisherman went straight back to the Mianus dealer and traded in his more sophisticated, new engine for another two-stroke.

Reports of the ability of boats like this soon spread. A 1907 article in *MotorBoat* described the experience of a fisherman in a 24-foot Hampton who found himself 15 miles from his home port on Bailey's Island in a developing northwesterly gale. "This boat made the run home," reported boat builder David P. Sinnett, "against the wind in two hours and forty minutes." By then, the fishermen had recognized the value of the gasoline engine, and it was estimated that only about a dozen of Casco Bay's Hampton boats were without an engine. It all meant more business for Mianus, but other brands, principally Palmer, Lathrop, and Bridgeport also began to appear.

By 1902, lobstermen were fitting their sloops or smaller boats with gasoline engines equipped with a clutch so that the engine could be put in neutral as pots were hauled. Suddenly, the labor of tending pots, 25 to 100 were then typical, was made a lot easier. At Eastport, Maine, where the first American sardines were canned in 1875, the fishermen began installing engines about 1900 or 1901. Within a few years, engines of 10 to 50 horsepower were common. "During the summer months it is claimed that there is not an hour of the day or night in Machias Bay that the *chug, chug,* of a motorboat cannot be heard," reported *MotorBoat* in 1910. Two decades later, in 1920, there were reported to be 1,000 power boats operating in Passamaquoddy Bay and neighboring waters.

By then, a variety of engines were in use burning gasoline, kerosene, and crude oil. So it went, all along the East Coast. On Long Island's Great South Bay, the big fleet of oyster boats began to have motors installed that drove both the boat and the dredge-hoist mechanism while, all along the island, all manner of skiffs and sharpies were converted to carry a single-cylinder engine. In Chesapeake Bay, the crabbing skiffs were increasingly found putt-putting in the creeks, rivers, and the Bay itself. By 1916, it was estimated that 25 percent of Maryland's crabbers had a gasoline engine in their boat and there were concerns that crab dredges would largely destroy the harvest. On the New Jersey shore, and at Cape Cod and Newport, the beamy, shallow-draft, Cape Cod catboat that was the staple inshore fishing boat, began to be converted to power by 1900 or 1901. Mianus and Lathrop were favored, and their installation in a catboat highlighted another advantage of such engines. Typically, they were mounted well aft so that the motor disturbed neither the centerboard nor the live-fish well. This meant that the engines were installed at a rather considerable

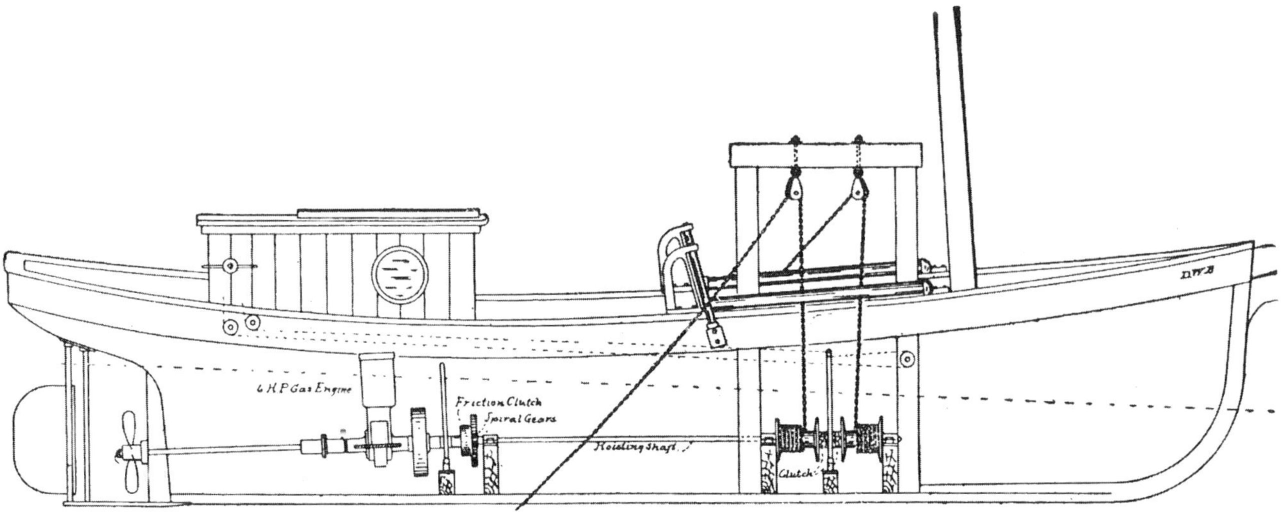

Oyster sloop, Great South Bay, Long Island, New York, circa 1917. This profile shows a 32' x 11' oyster sloop originally designed for sail but now fitted with a 6-hp, two-cycle engine. A double-drum hoister was built especially for the engine. It eliminated the need for a crew member. In fact, with the installation of the engine, this boat could be handled by one man.

angle, but the two-cycle machines were perfectly adapted to such mounting, for there was no requirement for a four-cycle's lubrication system.

In Boston itself, hub of so much of the region's fishing economy, the inshore fishing dories began to be converted when some fishermen, becalmed with a catch of haddock and cod near Boston Light, saw a motor launch speed past on its way into the harbor. According to an article on the Boston dory fleet published in *MotorBoat* in October 1908, the launch owner "was hunted up by the fishermen.... When the first motor fisherman got well into harness, her owners' wives began to appear on the streets in new dresses."

A power fishing dory may have appeared in Boston as early as the winter of 1898–'99, built by Gerry Emmons, whose boatshop in Swampscott, about 10 miles up the coast from Boston, had made its reputation for supplying the Swampscott dories used by a generation of recreational sailors. C. S. Wells, a designer who also lived in Swampscott, gave credit to a local fisherman and boatbuilder named Brackett for first conceiving the idea of building a round-sided dory as distinct from the straight-sided banks dory. He also claimed to have been the designer of the Emmons motor dory. Soon, a thriving business arose, with more boats built in Swampscott by W. H. Chamberlain and in Boston by A. W. Toppan. As it turned out, there was a big enough market to support all these builders and others besides. But Wells claimed that "more power dories are built in Swampscott and Marblehead than all the rest of the country combined."

He noted that the most successful boats had high-speed motors with carburetors rather than mixing vaporizers. "The vaporizer valves are the poorest mixers of all and, after a cold night, are generally found frozen solid and, when piped to take warm air off the exhaust piping, they must be thawed out before the motor can start, which process is the cause of nine-tenths of the fires on fishermen's launches."

The Boston dory fisherman population was largely composed of Italian immigrants who formed a colorful attraction on the waterfront. They bought Mianus and, later, Bridgeport, two-cycle engines by the dozen. Hartfords, Palmers, and Lathrops were also adopted and larger boats, 25–35 feet long, were powered by two-cylinder models producing up to 20 horsepower. Many of these boats were built by Willis J. Reid in East Boston, and they were not to be confused with dories. They were heavily built, with oak keels and ribs, and one-inch-thick planking. A 30-footer might have a 10- or 12-foot beam and draw 2½ feet. A cabin aft housed bunks and a stove. The engine, and the steering wheel, was mounted there on the forward-facing bulkhead. Thus, the helmsman steered with only his head showing, eliminating the possibility of being washed overboard. Commenting on boats like this, which he himself built in some quantity, A. W. Toppan said: "The dory type was not a handsome craft but on account of fine rough water qualities there arose a demand for this type for pleasure use." Toppan then phased out of supplying the fishermen.

It took really bad weather to keep these boats tied up. They ranged freely all over Massachusetts Bay, sometimes even voyaging to Maine, and the three or four men aboard each one — hand-lining for groundfish, cod, and halibut — earned a good wage. The boats, according to a 1906 *MotorBoat* article, lasted five or six years. "Engine care was seldom anything more than a bucket of salt water sluiced over them at the end of a day." In his 1910 article, George Story Hudson was less kind. "A considerable proportion of these fishermen are lamentably ignorant as to the construction of a gasolene motor. Some know that the switch, batteries, gasolene and lubricating oil are important factors in operating the power plant, but beyond that their practical knowledge is scant."

Aboard the Boston schooners, however, Hudson found that the many elements relating both to efficient and safe operation had been very well thought out. The big timbers of a schooner provided an ample place to locate the engine bearers, and cement was poured in beneath the engine to prevent the buildup of oil or gasoline vapors under the floor timbers. The stuffing box was mounted inside where it was readily accessible. Great ingenuity was displayed in the design and mounting of a muffler, for skippers didn't want to scare off swordfish and mackerel with the bark of their heavy-duty, two- or three-cylinder, two-cycle motors. According to George Story Hudson, a muffler as big as a nail keg and well integrated into the boat's machinery converted "the roar of a two-cycle motor to a lisping pant that wouldn't alarm a swordfish afflicted with the worst case of nerves you ever saw.... Even the location of the exhaust pipe is carefully considered, recent practice placing the outlet through the deck rail on stern or quarters, where it will not be flooded by following seas or choked when the craft is heeled to a wide angle and seas endeavor to slop into its sooty throat."

The great fear always had to do with explosion and fire. But within a few years of the first installations, this issue was apparently addressed to general satisfaction. According to Hudson, writing in 1910 of the Boston boats, no fire traceable to a vessel's engine had been recorded. By then, it was standard practice for larger schooners to have fuel tanks mounted on deck, just forward of the cabin trunk. The tanks, he wrote, "are constructed with great care and the piping to carburetors is of the best, with painstaking installation. These tanks do not become leaky, as a rule, though subjected to very hard usage from the rolling and tossing of the vessels. Should a tank burst, its contents would flow on deck with no possibility of getting below even through pipe holes in the trunk, which are so arranged that plugs can be inserted should any sign of leakage be apparent. In heavy weather the tanks are often submerged by boarding seas but fastenings as yet have not been wrenched out of place, at least no report is at hand of such a mishap."

The engine for John LaFond's Mackinaw boat was built in a shop in the back of a hardware store in Two Rivers, Wisconsin. The man who built it was William R. Kahlenberg. Together with his brother, Otto, William Kahlenberg had opened the machine shop in 1895 and soon was involved in the installation of machine tools and steam engines in area factories and on farms, and in converting simple steam engines to compound types. Although Two Rivers was a fishing community, steam engines were recognized as being impractical for the sturdy, open, 14- to 34-foot Mackinaw boats that, with their brown, green, and maroon gunwales, were a colorful sight at the town's wharves.

The rendition of William Kahlenberg's building of this engine for John LaFond became a part of Two Rivers lore as people recognized its creation, even at that time, to be an epoch-making event. Thus, a brief description of the first engine's creation was almost always included in later newspaper or magazine articles about the company that grew out of this engine, Kahlenberg Brothers. Not recorded, however, is how Will Kahlenberg, with a sound background in steam practice but no experience in gasoline engines, got his engine right on the first try. One can only assume that, since Will was both mechanically gifted and a good student, he followed engine development as reported in the press of that era. The sole detail regarding the first engine's inception tells us that Kahlenberg met with LaFond and that "out of this huddle emerged drawings, patterns, and eventually the engine itself." Years later, in a brief, 1924 retrospective on the company, Will Kahlenberg pointed with justifiable pride to the fact that "from the time of building the first engine, parts were always made with jigs and templates and therefore standardized and interchangeable; an owner ordering from some remote corner of the earth or at home was always sure of a part that would fit."

Kahlenberg's engine was a husky, single-cylinder, two-cycle with a make-and-break ignition system. It was a two-port engine. That is, the fuel/air mixture entered into the engine's base via a check valve, was compressed by the descending piston and forced through a transfer port to the cylinder. The LaFond engine was remembered by Will Kahlenberg as an eight-horsepower machine, and it probably developed peak power at around 400 rpm. With what proved to be great foresight, Kahlenberg equipped this first engine with a control lever that permitted the engine to be stopped and reversed, and it also adjusted the timing, which was important to speed control on a two-port engine. Thereafter, Kahlenberg brochures included the following paragraph:

William R. Kahlenberg (r.), 1871-1957. Looking dapper, Kahlenberg is seen here aboard a yacht fitted with one of his engines. (Courtesy, Sandra Zipperer)

> Reversing — the great obstacle that has confronted builders of marine gasoline and oil engines — is a simple accomplishment of the "Kahlenberg," and they can be reversed, without stopping, from full; speed ahead to full astern, just the same as any steam engine, and are just as positive. This is accomplished without the use of a gear of any kind in the construction of the machine, and it is the only reversing engine on the market that may be depended upon to work properly. We are aware of the fact that all jump spark engines will reverse on the spark, sometimes, but our method of reversing is entirely different, and is always positive.

LaFond's engine cost him $400 and the sight of a Mackinaw boat powering reliably out into Lake Michigan made a profound impact on all who witnessed it. Charles Davis, who visited Two Rivers in 1910 on assignment for *MotorBoat* magazine, wrote:

> "What the gasolene motor has done in helping the hard-working fisherman to wrest a living from the lakes is well-illustrated in that class of fishermen who set and haul gill nets in the deep cold waters of Lake Michigan. Where, in former years, they had to depend on Mackinaw sailboats to take them eight or ten miles out in the lake, and where the nets had to be hauled in by hand, a job that at times taxed three men, they now have a fleet of a dozen or more strong, beamy, excellent sea boats driven by gasolene motors, and the nets are all hauled by the same power which operates a peculiar net hoisting drum over a zinc roller, so that the crew of three, instead of six, can now all stand inside of the cabin and work, instead of having to stand on the deck of a wildly plunging and rolling boat. . . . Albert turned over the flywheel of the powerful two-cylinder, two-cycle Kahlenberg motor, and with the loud bark of her exhaust shooting through an iron pipe up out of the cabin roof, we started."

Almost immediately after the launching of John LaFond's boat, an engine was ordered for a tow boat, and another by a fisherman in Milwaukee. On the Great Lakes, in those early days and for years afterward, people called the engines not a Kahlenberg but a "Two Rivers." When Davis visited Two Rivers in 1910, he found, in that one town alone, about a dozen powered boats working summer and winter, "sometimes cutting their way through seven and eight inches of ice, for their best catches are made in cold weather." They had one Kahlenberg driving the boat and a smaller, single-cylinder rigged to operate the hoisting drum. It was a hard life, but it could support a fisherman and his family well, and it was not without the intrinsic charms noted by Charles Davis: "As we pitched out toward the rising sun in the crisp morning, with a fresh southerly breeze sweeping through the open windows and doors, as the light, buoyant hull rolled deck edge to in the short beam sea, the cabin was all aglow with morning sunlight and shadows."

Fishermen, commercial operators, and then well to-do yachtsmen in Chicago began buying so many engines that by 1907, a fine new brick and concrete factory had been built. Often, owners' experience with their Kahlenbergs, which were painted in a pleasing "steel blue" enamel that contrasted nicely with the highly polished brass fittings, wrote letters expressing their satisfaction. Said a Chicagoan who installed a two-cylinder model in his cabin cruiser in 1908: "I desire to say that your company builds one of the best marine gasoline engines in the country. This engine drives my boat, which is a very heavy craft, about eight miles

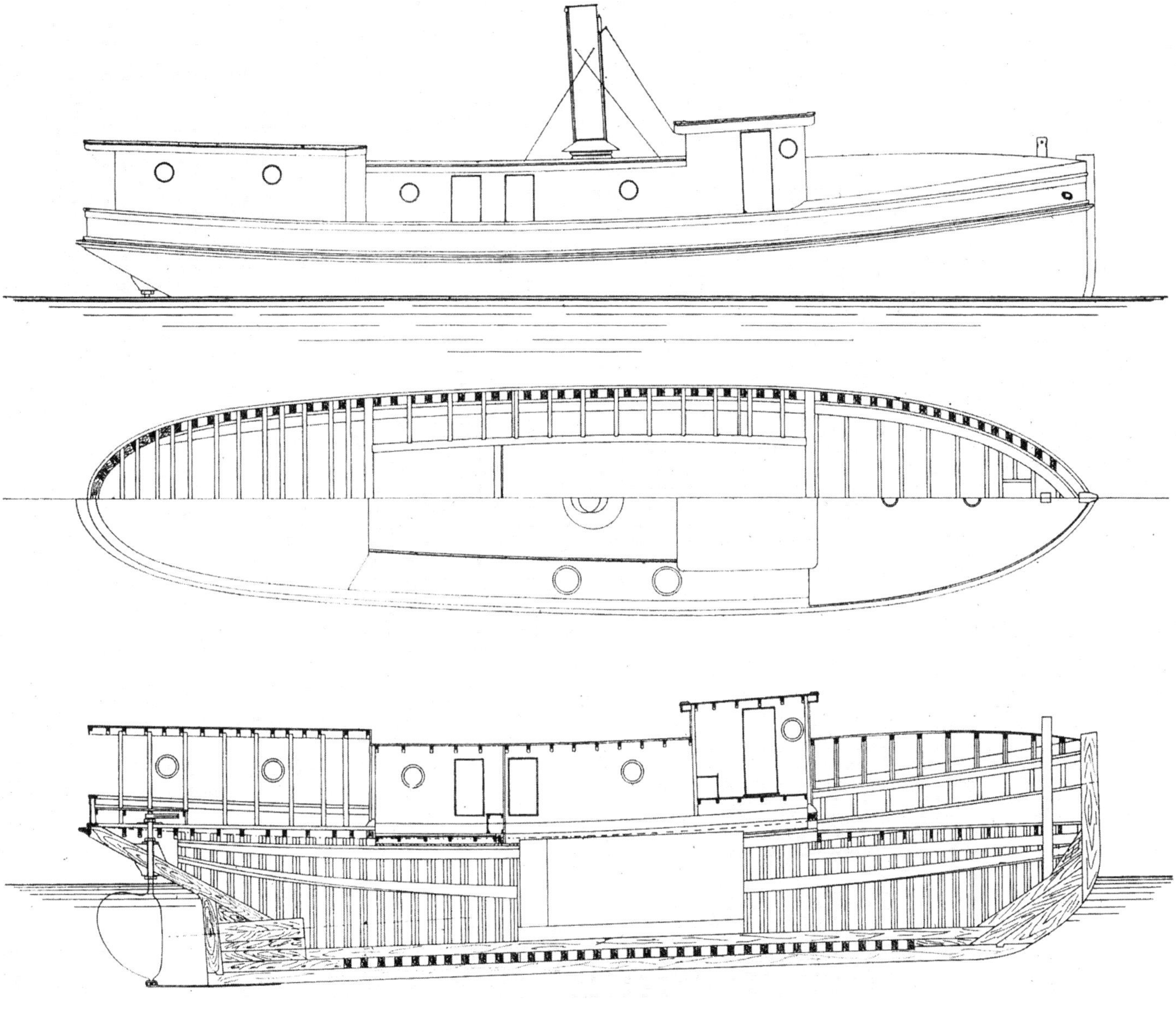

Great Lakes fishing vessel, 1918. Built by the Burger Boat Company in Manitowoc, Wisconsin, this "deep-sea fishing tug" measured 72' x 16 x 8' and was typical of its generation of gasoline- or steam-powered boats. The deck houses were designed to protect the crew from the elements, especially during late fall and early spring fishing.

Pound boat. This open boat was 40 feet long and was powered by an 18-24 hp Kahlenberg gasoline engine. (Courtesy, Sandra Zipperer)

an hour and gives the best of satisfaction. It uses about one gallon of gasoline an hour..." Soon, Kahlenbergs were a fixture in boats owned by members of the Chicago Yacht Club and the Club made Two Rivers a stop on its annual power boat cruise to Sturgeon Bay.

For the fisherman, however, it was the introduction by Kahlenberg of a semi-diesel engine that would have the greatest impact since the company's first gasoline model. Will Kahlenberg began experimenting with a single-cylinder, heavy-duty, oil engine in 1914. This engine, which was started on gasoline but then switched over to run on fuel oil, powered factory machinery for a full year before a marine model was built. It was sold to a pair of Two Rivers fishermen in 1916. Although the Kahlenbergs always referred to the new engine as a "semi-diesel" or "heavy-duty crude oil engine," there was the typical reference in the lay press to the engine as being a *diesel.* The 1916 catalog introduction to the engine made clear how the machine operated.

> "All engines of this type are started in the usual manner on either gasoline or distillate, and after the machine has been running two or three minutes, the fuel pumps are turned on and the engine then operates on the semi-Diesel principle — taking pure air only in the base, compressing it, and just before the piston reaches the top of its stroke, the fuel oil or kerosene is injected directly into the combustion chamber by the small injection pumps, one for each cylinder, timed exactly in rotation with the engine."

Beginning with Will Kahlenberg's first two-cycle engine and its built-in reversing mechanism, all the firm's literature stressed how easy these machines were to operate and maintain. (See Appendix for a first-hand account of operation.) It is for this reason, perhaps, that the company literature didn't go into great technical detail on the principles of that operation. But all these Kahlenbergs, whether gasoline or semi-diesel, were two-cycle engines. They remained so in 1922 when a new generation of engines was introduced. These were heavy-duty, crude-oil engines, "The New Kahlenberg Oil Engine." It was a hot-bulb engine with mechanically injected fuel. Gone was the previous model's ignition system used for starting. It was replaced with a torch that heated the hot tube — "starting tube" — within two or three minutes. Then, said Kahlenberg, turn on the air and fuel, "Pull the air throttle and the engine is agoing." The flames of the torches that were built into the cylinder heads to heat the starting tubes were ducted outside the engine compartment by a steel manifold. The new engine was available in two-, three-, and four-cylinder models of 20 to 200 horsepower, and the catalog showed the line drawing of a Pacific Coast purse seiner with a three-cylinder Kahlenberg in the engine room.

Kahlenberg

"SINCLAIR"

ANOTHER new Kahlenberg oil engine powered work boat which is unique among harbor craft is the floating service station "Sinclair", built by the Burger Boat Company for the Sinclair Refining Co. It is used for delivering fuel, lubricating oil, and ice to boats in Chicago yacht harbors.

The "Sinclair" is 54′ x 15′ x 5′-6″ and has a four cylinder 60-70 H.P. Kahlenberg oil engine which is controlled from the pilot house. The flexibility of this engine and ease of control make it especially desirable for a boat of this type where a great deal of maneuvering is required and many starts and stops are made.

KAHLENBERG BROS. CO.

Manufacturers

TWO RIVERS, WISCONSIN, U. S. A.

The 60-70 H. P. Kahlenberg Oil Engine Installed in the "Sinclair"

It was company practice to build each engine to a very high standard and then test it before shipment. Those who witnessed these tests reported that no engine left the factory until it was correct in every aspect. Sometimes, even after an engine had passed all its tests, Will Kahlenberg might give it his own personal tuneup. Said one owner: "This personal final inspection and regard for the individuality of the engine is refreshing in these days of quantity production, where parts are rushed through operations and inspections, and assembled and shot into the market with all possible speed."

The Kahlenberg catalog, of course, featured numerous testimonials. "It runs like a clock and I believe it is the best engine on Puget Sound," wrote Captain Nick Milosevich. From Captain John Dahlmer, whose 110-foot Gloucester schooner was powered with a four-cylinder, 100-horsepower Kahlenberg came, "I am well-pleased with the engine in all respects." Other testimonials arrived from the Great Lakes area from the Philippines, and from Spain. Will Kahlenberg kept them all, just as he kept records of each engine the company built and shipped.

The Kahlenberg brothers were early advertisers in *MotorBoat* magazine and, by the mid-'20s, these ads had helped to generate an export business. There were some years, in fact, that company files revealed higher export than domestic sales. On the Great Lakes, by the mid-'20s, it was said that there were more fishing boats powered by Kahlenberg than with all other makes combined. Demand was such that the factory was doubled in size in 1924, its machinery driven, of course, by one of the new, semi-diesels. The economy and easy operation of these Kahlenbergs made the earlier, gasoline models just about obsolete. The company, of course, still had the tooling for those engines. Kahlenberg catalogs, for a number of years after the new model semi-diesel was introduced, retained an offer to build the gasoline engine for any customer requesting one. At least as of 1924, a few such special orders were received.

Like its predecessors, the new models were robust. Kahlenberg was known locally for being a "crank about crankshafts," and the dimensions of the crankshaft journals and of the main bearings were impressive. A four-cylinder, 100-horsepower Kahlenberg had end bearings that were 10 inches long, the bearing shells cast of bronze with a layer of white metal lining that was bored to an exact fit. These 10-inch-diameter bearings weighed 17 pounds apiece. The company liked to point out to customers that mass-produced engines — and one can pretty well assume they were referring to Fairbanks-Morse — had iron-shell main bearings with poured-in babbitt. Total main-bearing surface on the four-cylinder was, the company calculated, over *4½ square feet*. That's probably why these bearings, although replaceable in less than a half-hour, so seldom needed replacement. Stories abounded of their long life. One owner ran his Kahlenberg for 16 years without a bearing replacement; another engine was run 21,000 hours and was still on its original bearings.

Such engines always found a ready market. Production continued steadily through the '30s, when Kahlenberg made some inroads into the towboats built for use on inland rivers. Export business continued right up until World War II, the last shipment to the South Pacific involving engines destined for fishing boats in Dutch Indonesia. Kahlenberg papers found by researcher Sandra Zipperer included the following regarding these particular engines.

> In 1941, just before the Japanese entered Java, the Dutch Indies, the Americans in passing through found five Kahlenberg 150-horsepower, three-cylinder units in crates on the docks there. These engines had originally been intended for Dutch trawlers. The Americans immediately loaded these on board ship for Australia and installed

> them in Australian sailing schooners, which were used by MacArthur's men to return to the Philippines in early 1942 to smuggle small arms and ammunition to the guerrillas. These and other similar boats played a very large part in the early campaigns against the Japanese in the Solomons and New Guinea. The results of the dependable performance on the 2000-mile non-stop run to the Philippines from Australia are reflected in the continuous fight waged by the American-led guerrilla warriors in the Philippines who carried on until MacArthur himself could return in proper force.

It was said that MacArthur himself was so impressed by these Kahlenbergs that he sent a personal envoy to Washington to request production of 300 engines, an order that was soon increased. Although the Kahlenberg factory had been increased in capacity, the government did question whether such a production level was practical. Will Kahlenberg, of course, saw little problem and promptly arranged to devote the plant's entire output to the war effort. "At the height of production," according to a paper written on the company by Sandra Zipperer, "the Company had over 100 employees and worked 70 hours a week and more to meet production needs. But despite quotas, no engine left the plant unless it met Kahlenberg standards." On September 14, 1944, Kahlenberg was awarded the Army-Navy Production Award and could then fly the special "E" award flag at the Two Rivers factory.

William R. Kahlenberg accepts the Army Navy E award. Kahlenberg's contribution to the war effort was recognized in this ceremony held at Two Rivers, Wisconsin, on Sept. 14, 1944. (Courtesy, Sandra Zipperer)

Like virtually every other marine engine builder, Kahlenberg faced a new world in the aftermath of the war. Export sales continued, thanks in part to the Marshall plan, but the domestic business fell. A new, six-cylinder, turbo-charged engine developed by Will Kahlenberg's son, Roger, in 1954 was not enough to stem this tide. It was a 34,000-pound engine that developed 600-900 horsepower at 300-400 rpm and it sold for about $60,000 but cost $90,000 to build. The era of the lightweight, high-speed diesel had arrived, replete with war surplus Gray Marine diesels, and Kahlenberg ceased engine production to focus instead on the air horns and other accessories for which it remains well known.

Within a few years of the time when the Kahlenbergs introduced their first semi-diesel in 1916, the conversion of fishing boats to power throughout America, not to mention Canada, was a done deal. Those first two decades of the 20th century brought with them the most profound changes ever wrought in this ancient trade. On Lake Erie, by 1919, 200 power boats were in operation, most in the lake's western portions around Sandusky, Huron, Port Clinton, Ashtabula, Fairport, Cleveland, Toledo, Erie, and Dunkirk, New York. They were powered by a variety of engines that included Scripps, Regal, Buffalo, Clay, and Peerless. These boats formed the nucleus of an industry that represented almost $1.3 million spent on wages, and the hard costs of operating and maintaining the boats. The boats, said to be worth an average of $1,000 each, averaged 29 feet in length and carried a three-man crew, half that of the steam tugs that also fished the lake.

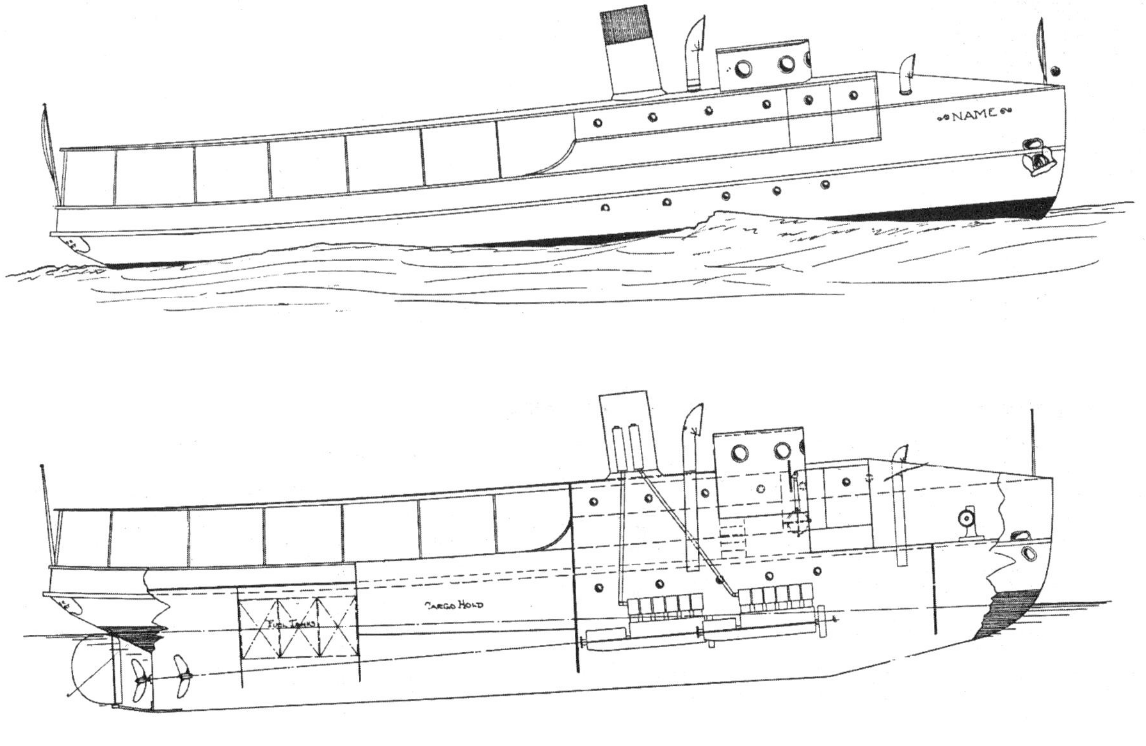

High-speed Great Lakes fishing boat, 1912. Built of steel and powered by three Buffalo gasoline engines, this vessel was designed for the Ranney Fish Company of Cleveland by Whittelsey & Whittelsey of New York. The vessel measured 80' x 16' x 6'. The engines produced 75 hp each and exhausted through the funnel. She was said to be capable of operating safely on any one of her three heavy-duty engines.

By the time the powered fishing boat had matured on the Lakes, it had also become a staple on the rivers. Along the Mississippi around Muscatine, Iowa, the clam fishery had grown into a mighty industry that supplied the shells used to make pearl buttons. From four to a dozen buttons could be cut out of each half shell. This fishery was not limited to the Mississippi. It flourished on the Illinois, the Tennessee, the Ohio below Cincinnati, the Arkansas, the White River, and others, too. Fleets of square-ended john boats, usually about 18 feet long, dragged the rivers. In the early days, around 1906 or '07, these boats were almost invariably powered by single-cylinder, two-cycle engines, most often a Ferro or Gray, but other brands were also used. Later, four-cycle models were installed by some fishermen seeking improved fuel efficiency. In a 1918 article in *Power Boating,* these boats were described as follows:

> The engines usually are installed in a rough-and-ready but substantial fashion. In many cases the exhaust pipe is simply run over the gunwale and the batteries and other ignition equipment housed in a small box installed abaft at the engine. A gasoline tank, which holds about 15 gallons, is carried in front under the forward deck, which is removable. The boat is fitted with an athwart-ship seat at the stern, with another, with rowlocks handy, forward of the engine. The skipper usually sits on this seat with the engine controls at one hand and the steering lever at the other. The latter is simply a lever on the port side of which the tiller ropes are attached. Some of these boats are painted in a gay and glorious fashion, with more regard for striking effects than artistic harmony. By and large, however, they are sturdy little craft and it is remarkable how much one of them will carry.

Clammers, Iowa, circa 1910. It was a calm day on the Mississippi River when these fishermen posed by one of the john boats used to drag for clams, and a gasoline-powered tender. Such craft were the basis for the industry that manufactured pearl buttons. (Courtesy, Iowa State Historical Society)

Charles G. Davis made a trip out to Peoria for *Motor Boat* magazine in the summer of 1910 and spent a day aboard one of these boats, a 20-footer. He reported that the fisherman:

> "Threw in his switch, primed his little single-cylinder Ferro motor, turned her over and she started first crack. Then, throwing out the switch, we shoved the scow out between the boathouses, clear of ropes and again the motor fired first crack and we were off, gliding out across the smooth river at about a six-knot clip.... he had the motor handy right between his knees so that he could tend the oil cups with one hand and steer with the lever on his left to which the tiller rope that ran clear around the cockpit through screw eyes and corner sheaves was made fast.
>
> "There was very little vibration....His motor cost him seventy-five dollars and his hull forty-five. He installed the motor himself. By setting his motor level and then raking his shaft with a knuckle joint connection, it shortened everything up. She didn't make much noise as she had a muffler on the exhaust pipe and her circulating water went right out the side through a small brass pipe."

Pearl fisherman, Illinois River near Peoria, 1910. Here we see a fisherman at the helm of a 20-foot scow powered by a single-cylinder, two-cycle Ferro engine that cost $75. The dry cell batteries used for ignition were mounted in the box on which the fisherman sat. A 10-gallon fuel tank was located in the bow. Relatively few clams have been hooked during the last drag but at least the fisherman didn't have to row. The pipe with its hooks was removed from the brackets, dropped over the side and dragged for 10 or 12 minutes. Then the engine was stopped, the drag hauled up and the clams or oysters were removed.

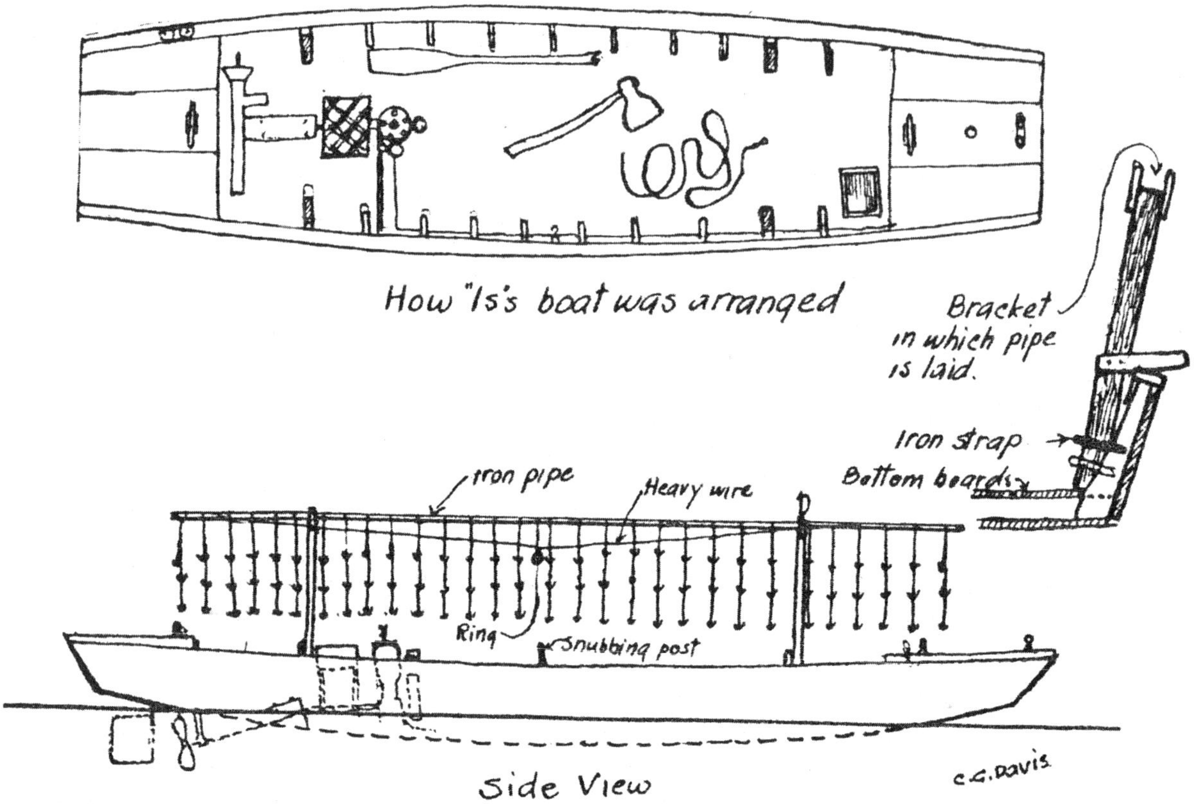

These sketches by C. G. Davis show the equipment of a fresh water pearl boat.

As was typical, the boat's 10-gallon fuel tank was located under the deck forward. Ignition current was supplied by four dry cell batteries and a coil contained in a box beneath the helmsman's seat. "As in many other industries," Davis concluded, "the internal combustion motor has made life easier for these clam fishermen. As I travel about the country, I find motors employed in so many kinds of work that I wonder how we ever got along without them."

Down America's great river system, where the Mississippi empties into the Gulf of Mexico, internal combustion engines were adopted by fishermen all along the coast. "The fleet that comes home to Galveston laden with tons of succulent red snapper to be shipped to the inland towns and cities has been making schedules of record regularity since the power of marine motors was added to the power of the wind," reported *Motor Boat* in August, 1914. "There are dozens of old-time fishermen in Galveston, too, and the 'chuff-chuff' of their motors is a constant sound up and down the channel. . . ."

Those who lived in south Louisiana around the turn of the century were witness to the entire evolution of fishing there, among the most dramatic changes occurring in the oyster fishery. "I can remember the day when the first gasoline invader entered the oyster fleet," wrote Herbert L. Johnson in 1909. "Oystering was then carried on almost entirely by Italians and Sicilians in their 'luggers,' shallow-draft boats ranging from thirty to forty feet long and rigged with one mast, carrying a huge lug sail.... With the advancing cupidity of man, and the introduction of steam, in smaller boats, the steam dredger invaded the field... scraping the bottom bare of everything.... Then the legislature stepped in, and the steam dredger was pronounced an outlaw."

Enforcing the law was a fleet of gasoline-powered yachts converted for their new application with a one-pounder on the foredeck. The gasoline boats immediately proved their worth, carrying roughly 50 percent more oysters than could the sailing luggers. The first boat was a conversion, the square stern cut away and replaced by a longer fantail. Reported Johnson: "On its deck was a low cabin trunk from inside of which came the undeniable noise which accompanied the big, slow-moving gasoline engine.... From that day on the conversion began."

There were smaller oystermen, too, that led a rougher life along the great expanse of marsh grass that verged upon the sea.

> "Probably the severest test to which the gasoline motor can be put is in the little, open boats used by the individual oystermen [who fish the coast reefs that] are lonesome places at their best — fringing the edges of the apparently limitless areas of waving salt marsh grass, miles and miles away from civilization, and even solid ground long distances apart, except straight down a few feet.
>
> "In fair weather and foul, these oystermen are at work in their poorly constructed boats, with barely the necessities of life.... Theirs is a very hard life, but the gas engine has done more towards making things worthwhile for them than any one invention in centuries."

Before the United States entered World War I, it was estimated that the Gulf Coast fisheries stretching from Brownsville, Texas to Tarpon Springs, Florida were producing $7 million worth of seafood annually and this jumped to more than $10 million by mid-1918. Oyster gathering, trawling, and shrimp seining were the backbones of this great industry. Like many others, John "Bill" Kosmich, who was born in 1909 and worked out of Biloxi, fished for both shrimp and oysters. "It was seasonal," he said. "We would shrimp until Christmas and then, after January 1st, we'd go for oysters. At first, we sold them by the hundreds, then by the gallon. But you had to shuck 100 oysters to make a dime. We'd oyster until May lst and then do more shrimping."

Much of Kosmich's fishing was done aboard Biloxi schooners and Biloxi luggers. The schooners remain, in the mind of many, the signature vessel of the Gulf Coast. Built of Louisiana cypress fastened with galvanized nails, the boats were sometimes 100 feet in length. Painted white or sometimes gray, the schooners' spars were scraped and coated with tallow. The boats were served by two or three luggers, some 30 to 40 feet in length and, when the first engines began to be installed, they tended to be two- or three-cylinder Lathrops of 24 to 30 horsepower. "A Lathrop lasted a good long while," remembered Kosmich. "You couldn't do much to hurt them. You could run 'em hot, for example, which you'd never do with the high-tech engines of today."

Biloxi lugger Lillian Malissie, *circa 1930. It is rare to see a lugger together with the giant seine nets used for shrimping. This photograph was taken by Mary Ethel Disinukes under the Depression-era program sponsored by the Works Progress Administration (WPA). The location and low height of the deckhouse suggests this boat had been converted from a sailing lugger. (Courtesy, Maritime & Seafood Industry Museum, Biloxi, Mississippi)*

Kosmich remembered the entire progression of internal combustion within the oyster fleet. The Lathrops were soon joined by another product of Connecticut that found widespread acceptance by fishermen, the Palmer. The old oyster schooners were the foundation of a way of life that, if difficult, is remembered with some nostalgia by those who lived it. "A good fisherman got by," said Kosmich. "You had gardens, pigs, and cows, and you could always get crab to eat or shrimp. It was a lot cleaner in those days, too."

A big oyster dredge, such as the 56-foot *Ruth J.* operated by the Louisiana Oyster & Fish Company of Berwick, Louisiana, was a substantial boat that could make nine miles per hour powered by a 50-horsepower Globe engine. The average horsepower reported for dredging or tonging of oysters was reported to be about 20 horsepower; for shrimp trawling 12 to 20 horsepower. On the biggest schooners used for seining, a seine boat was carried, usually powered by a five- or six-horsepower Frisbie or an eight- to 10-horsepower Clay, although boats with Automatics or 'Frisco Standards were also in evidence. Reporting for *Power Boating* in 1918, Harry Dunn noted that the "fishermen with whom the writer has talked seem to believe that the Frisbie, however, is the engine for the small schooners, inasmuch, as they claim, it uses less oil for the work

The last schooner race, July 4, 1938. The absence of topmasts indicates that these Biloxi schooners have been motorized. No longer did they need a lofty rig to speed reliably to and from the oyster beds. But they did retain two masts. After the Mississippi legislature made it legal to drag for oysters under power in 1933, many schooners had their main mast removed and a bigger engine installed beneath a house. (Courtesy, Maritime & Seafood Industry Museum, Biloxi, Mississippi)

done, and is more economical when used in connection with the sails. He reported, however, that four new schooners built by the Sea Food Company of Biloxi were all equipped with heavy-duty Clays, an engine that found favor with fishermen from the Midwest to the West Coast.

These Clays were interesting machines. They owed their design and manufacture to an inventive machinist named Edward Henry Clay of Cleveland, Ohio. In 1898, Clay, with the help of his brother, Cassius, constructed a four-cycle, overhead valve engine for his own use. As sometimes happened in those early days, several of Clay's neighbors were impressed enough by the engine's performance to ask him to build one for them. This was the foundation for a company, whose focus was always the fisherman and whose motto, "The Honest Clay," expressed a message of dependability and economy. At some point in his company's development, Edward Henry Clay began using the image of the American patriot, Henry Clay, as a symbol of the firm's integrity. In ads, the company referred to "the statesman whose integrity proved a symbol for public life and endeared him to the hearts of the people. We have not chosen the name HONEST CLAY for our engines without full realization that we must measure up to the ideal incorporated in the words."

2-Cylinder
40-50 H. P.
Model

HONEST CLAY
HEAVY DUTY

EVERY American citizen knows the story of Henry Clay, the statesman whose integrity proved a symbol for public life and endeared him to the hearts of the people.

We have not chosen the name HONEST CLAY for our engines without full realization that we must measure up to the ideal incorporated in the words.

HONEST CLAY engines must be RIGHT. We would rather build one engine a year and know that it was *right* than to build 5,000 engines of questionable construction.

For twenty years HONEST CLAY engines have run true to the claims made for them. Fishing and commercial craft on the Pacific Coast powered with these engines have demonstrated year after year their durability, serviceability and economy of operation.

The HONEST CLAY engine represents a genuine economy in first cost, operating cost and comparative service. Any Clay engine owner will confirm this statement.

They will operate efficiently 365 days in the year, on gasoline, kerosene or distillate—and stay on the job every hour.

We have made some changes in the manufacture of Clay engines, but we will not vary from the underlying design which has made this engine so great a success. The changes represent refinements in construction, details which add to the general efficiency and utility of the HONEST CLAY.

For the benefit of your good business judgment, we ask a careful investigation for the Clay engine. Go over it thoroughly. Estimate every possible feature. Be convinced of its simplicity, the accessibility of its parts, its extreme flexibility.

No matter what the size of your boat, what your requirements may be, there is a model of the HONEST CLAY to fit the need. Clay engines are made in sizes from 6 h. p. to 100 h. p.—single cylinder 6, 8 and 10 h. p.; double cylinder 12-14, 16-18, 20-22, 40-50 h. p.; four cylinder 25, 35, 50 and 80-100 h. p.—all four cycle, heavy duty engines.

KEEP THIS IN MIND—HONEST CLAY engines are *moderate* in price—but you cannot obtain a more reliable engine or one that will give greater satisfaction even should you pay much more for another make.

Enlarged manufacturing facilities—five times the capacity of our plant two years ago—make immediate delivery possible.

Call on nearest representative for demonstration or write for our complete illustrated catalogues. Don't delay.

The Clay Engine Manufacturing Company

764 East 72nd St., CLEVELAND, OHIO

WESTERN DISTRIBUTORS:

PACIFIC MARINE ENGINE COMPANY, 78 Marion St., SEATTLE, U. S. A.

OREGON: Evinrude Motor Company, 211 Morrison St., Portland

BRITISH COLUMBIA: A. W. LePage Gasoline Engine & Supply Co., 1773 Georgia St., Vancouver; Imperial Machine Works, Prince Rupert

The Clays were all heavy-duty types with crankshaft and connecting rods machined from carbon steel forgings, cams made of cold-rolled steel, and valves whose cast-iron heads were threaded and welded to cold-rolled steel stems. By about 1917, the engines had progressed so that they were not only overhead valve, but overhead cam as well. This valve mechanism was arranged so that the cylinder heads could be removed without affecting it. A pressure lubrication system was standard and, as a practical matter, "All nuts, bolts, studs, and fittings are standard and can be purchased at any hardware store." This entire approach to design and manufacture was apparently quite effective for, when Edward Clay made a trip to the Gulf Coast in the 1950s, he quite accidentally stumbled across one of the machines still performing daily service in a fishing boat.

The Clays sold their company in 1917. The name was changed from the Clay Engine Company to the Clay Engine Manufacturing Company whose president, Horace Newsome, reported a capitalization of $100,000. The model line grew steadily until, by 1919, it included models ranging from four to 100 horsepower. By then, too, Clay's reputation was so well established that the firm was able to move to a larger factory. This happened to be the then-empty factory in Cleveland that had been occupied by Harry A. Lozier's automobile company. Production of the V-12-powered HAL had commenced in 1916 but things never really developed as planned, and Clay's new owners bought or leased the plant in 1918. Subsequently, it acquired even more space so that, in 1922, it claimed to have "five times the capacity of our plant two years ago."

As the '20s wore on, the company entered into the manufacture of a variety of equipment, including chain hoists and stationary engines and it also began taking in machining jobs. None of this was enough to keep Clay going during the Depression and its operations ceased in 1930 or '31. Edward Clay continued to work for much of his life as an inventor and engineer. He died in Cleveland in July 1960.

In Cajun country, on the west bank of the Mississippi, there existed a world that was quite distinct from that of the New Orleans fisherman. Great pride was taken in the construction of sailing luggers and, later, fantail stern, powered boats. These boats were generally built of old-growth cypress, magnificent trees that rivaled in age and grandeur the California redwoods. Back in the Louisiana bayous, the need was not for heavy-duty engines but for reliable, smaller machines that, by around 1904 or '05, began to be used in a boat that replaced the region's skiffs. This "was the bateau, also commonly called 'gas-boat,' 'joe-boat,' and 'john boat,'" recorded Malcolm L. Comeaux in his *Atchafalaya Swamp Life*. "It is a long, heavy, flat-bottomed boat, with a blunt bow and stern, steered with a rudder and propelled by a single-cylinder inboard motor. It is not known when the first bateaux were used in the basin, but a 'gasboat' was advertised for sale in the *Morgan City Review* in 1907 and a boat whose description fits that of bateau was advertised for sale in 1908."

These boats, heavily built of cypress, were powered, often as not, by one- or two-cylinder, two-cycle motors built in Jackson, Michigan by the Lockwood-Ash Motor Company. The company had been founded in 1901 as Lockwood Brothers by Fred and Arthur Lockwood who began repairing bicycles and electrical equipment in a 13 by 30-foot building on East Courtland Street. By 1903, they had become dealers for Pope-Waverly and Oldsmobile, the latter an especially lucrative franchise. Within just a few years, in 1906, the Olds magazine, *Chug*, would report an anticipated $2 million in orders.

The increase in business resulted, in 1906, in the Lockwood brothers moving to larger quarters. They also took in a partner, named W. L. Ash, who had a business making spark plugs. Now, as the Lockwood-Ash Motor Company, the firm introduced two small marine engines. These motors were an outgrowth of experiments conducted by the Lockwood brothers beginning around 1904 or 1905. Within the family, precise dating of events was never particularly detailed, but the brothers did sell their interest in the garage in 1907 to focus on the marine engine business.

The lineup contained a single-cylinder two-horsepower, and a two-cylinder, five-horsepower, jump-spark motor. These were soon joined by the four-cylinder, 10-horsepower Model D. These little motors, produced in a wood-framed factory building on the corner of East Main and Horton Streets in Jackson, found widespread acceptance. In 1912, the company reported production of 8,000 motors. Sears-Roebuck sold them as the Motor-Go. How long this affiliation lasted is not recorded but, in December 1923, the company's then-president, A. L. Lockwood wrote a letter to marine engine dealers stating a sales policy that sounded quite a bit like what Sears might have coined.

> "In 1924 we are going to sell more than L-A Marine Engines — we are going to sell marine engine satisfaction. And we are going to sell it on a Golden Rule basis. All so-called free trial offers, all of the usual money-back guarantees we are going to sweep aside. They have been used so much, and with so many jokers and strings, that they have come to mean absolutely nothing. We simply want marine engine dealers to know, and customers to know, that from now on, L-A Marine Engines must give satisfaction to their owners, or the Lockwood-Ash Motor Company will make good.... That's all there is to it — no conditions, no strings."

A variety of refinements and new models followed the introduction of the first L-A motors, including the Model 44, a four-cylinder, four-cycle engine of 16-20 horsepower, around 1918. By then, however, the company was focusing equal attention on outboard motors, the first of which it had offered in 1914. When W. L. Ash died during the winter of 1921, his son had no interest in continuing in the company. The Ash family sold its stock to the Lockwoods and the company name reverted to its original Lockwood Brothers. Ultimately, the focus on outboards grew and Lockwood became a familiar name on the racing circuit. Ultimately, this led, in 1929 to the merger of Lockwood Brothers with the Outboard Marine Corporation (OMC) and the closure of the Jackson factory. The Lockwood Division of OMC, however, was short-lived. It fell victim to the Depression.

All this did not mean the end of the L-A marine engine. The Nadler Foundry and Machine Company, a Lockwood dealer in Plaquemine, Louisiana, purchased the rights and patterns and began manufacturing the engines. They were now known as the Nadler. It was a business that would continue for some 20 years. Looking back on this enterprise in 1974, the company's then–vice president, J. Allan Nadler, Jr., recalled that the engines "were produced in somewhat limited quantities for approximately 20 years and were used primarily by the loggers and fishermen in South Louisiana."

These old Lockwoods and Nadlers lasted a long time, although they were often run nearly year-round. They powered the fishing boat during the season, when catfish, gar, buffalo, and other fish were caught on hook nets. This was important work for Cajun fishermen and others in the swamplands for, in the general absence of cash, fish were typically a key item of barter. In the winters, the engine might be removed from the hull and rigged to cut cordwood. In the late 1970s, collector Robert Mayeux took his houseboat for a trip into Bayou Jake where he found, in a sunken bateau, a mud-covered, two-cylinder, eight-horsepower Lockwood-Ash built in 1918. Later on the same trip, he found an old fisherman *still using* an eight-horsepower Nadler-built model made in 1928. "The old fisherman said he had been fishing in the bayou all his life and had only praise for his old, rusty, well-used, fishy-smelling engine," Mayeux recalled of this meeting. "He said that he tried those new outboards, but being a commercial fisherman, it wore out in just one year. So he vowed to keep his old, reliable engine and would not think of selling or trading it for all the fish in the bayou."

Other engines were used in the bayous, too. Fairbanks-Morse two-cycles still turn up on occasion. Old-timers remembered that the fastest boat on Bayou Terrebonne in the early 1900s was the 32-foot long, five-foot-beam *Fentress,* powered by a two-cylinder, 12-horsepower Fairbanks engine. Some of these Fairbanks engines were marine models, others cleverly converted from stationary versions. Such conversions were common in South Louisiana, and a stationary engine modified for marine use entered the lexicon as a "Felequa" — pronounced *fee-lee-qua* or *fee-lee-quak.* "You won't find the word in a dictionary," said engine collector Dick Gibbens, "but all along the Louisiana coast, you'll hear it used in reference to an old vertical engine." Vertical was key, because folklore suggests that the word may have evolved from a French word for a kind of "bird that stands erect." Gibbens's Felequa may be considered a classic example of the type. It is an International Harvester model known as the "Famous Vertical." When converted to marine use, such engines had one flywheel removed and their hit-or-miss governor replaced by one that permitted throttle control. These felequas were commonly found in a variety of oyster luggers used in tidal waters.

As self-sufficient as fishermen anywhere, the Cajuns and Creoles in Louisiana were wary of new ideas involving engines. It was not a fisherman but the head of one of the largest shrimp-packing firms in New Orleans who responded to a letter from a manufacturer asking that he consider installing a new sort of engine in his shrimp boats. The letter came from Columbus, Indiana, and had been written by Clessie Cummins. By 1922, Cummins had what he believed was a practical, single-cylinder diesel and he persuaded the shrimp packer to permit installation of the Cummins in a local boat. The goal was to demonstrate that the Cummins-powered boat would outperform a comparable gasoline-powered vessel.

There were difficulties from the outset as Cummins recorded in his *My Days with the Diesel.* "I could not convince the boat crewmen that our higher speed engine, operating at 600 to 800 rpm, required a different kind of propeller. With the gasoline engine our unit was replacing, these men had been using a propeller about twenty-two inches in diameter with a twenty-two inch pitch. . . . Because of the increased revolutions, our engine needed a propeller of twenty-two-inch diameter and only thirteen- or fourteen-inch pitch. However, the crew refused to leave the dock with that propeller."

Eventually, Cummins tricked the crew into believing that their boat's new propeller was similar to the old one and the tests began. Once it was learned that the gasoline-powered boat was towing a shrimp trawl 20 feet shorter than Cummins' boat, the diesel-powered craft demonstrated much greater efficiency than its competitor. This resulted in orders for six or seven engines and Cummins returned to Indiana believing his company was well on its way to success. This proved to be wishful thinking, however. Not only did the engine's injectors require cleaning every three or four hours, but then the counterweights attached to the crankshaft began flying off, injuring crewmen and damaging boats. This sent Cummins back to the drawing board. Several years later, when Cummins had a new model ready for market, he looked elsewhere than Louisiana for sales. After World War II, when diesels came to the Gulf Coast fishing boats in a big way, they did so as GM models together with some converted Caterpillars.

Just as the internal combustion engine had a major impact on fishing elsewhere in the country, it had also brought profound change to fishermen on the West Coast. As early as 1900, a reporter for *The Rudder* wrote of what were then being referred to in San Francisco as "perfume boats."

"I am astonished to see how many of them there are.... Now they are beginning to be the rule and not the exception; some carry a big sail and using the engine as an auxiliary, and some discarding the sail entirely and trusting to luck for a tow. Yes they blow up, and get on fire, and sink, and break down, and raise the devil generally. But they are getting more and more practical every day...."

While the fishermen were early adopters of the new technology, they tended to follow the lead of what worked well for colleagues. Among the gasoline engines that achieved early popularity in San Francisco and then maintained it was the Hicks and the 'Frisco Standard. The Hicks, in particular, found a loyal core market among San Francisco's Italian fishermen, who began converting their colorful, lateen-rigged feluccas to power. In 1960, Hans R. Buckwalter recorded his own personal memories of these boats in an article for *Yachting.*

"Today, these boats are powered with almost anything and everything in the way of marine motors," he recalled, "but two or three decades ago no engine was considered right for the boats except the old slow-running West Coast–type one-lunger; that faithful combination of ballast [He recalled an early four- horse-power Hicks as weighing 800 pounds.] and engine with its great flywheel." He recalled the Hicks as being preferred "almost to the point that no other brand would be accepted."

Buckwalter described the starting drill for one of these engines as follows: "The fisherman lifts the oil cloth carefully from the top of the engine, solemnly takes the oil can full of gas and squirts it in the little cup on the top. Then he opens the exhaust valves. Now the great flywheel is turned, its momentum carrying it on and on. The man steps around the engine and lets down the valve lifts. There is a wheeze and a chug-chug. The water is turned on, with its wonderful soothing effect. The lines are cast off and the vessel moves away over the water seemingly without effort."

Thus began the evolution of the traditional felucca. The deck was opened to permit engine installation, a shaft alley bored and a section cut out of the deadwood. Soon, a small engine box was erected to protect the exposed cylinder head. Buckwalter recalled that it was about 1915 when the plumb felucca bow was changed to a clipper bow with lots of flare to better handle the chop outside the Golden Gate bridge. Then the stern began to be altered to a rounded or canoe-style. By the '20s, small pilot houses began to be built around the engine. They had big windows port and starboard but were open fore and aft to permit plenty of engine ventilation. Thus was born the boat that became famous as the Monterey Clipper.

As for the Hicks company itself, it was begun in San Francisco by James L. Hicks in 1889 to import and sell sewing machines. Hicks family lore suggests that James Hicks was a skilled tool and die maker whose interest was rather naturally attracted to the gasoline engines on the city's waterfront. It is certain that Hicks's focus changed to engines because by 1892, his business was listed as J. L. Hicks Machine Tools and

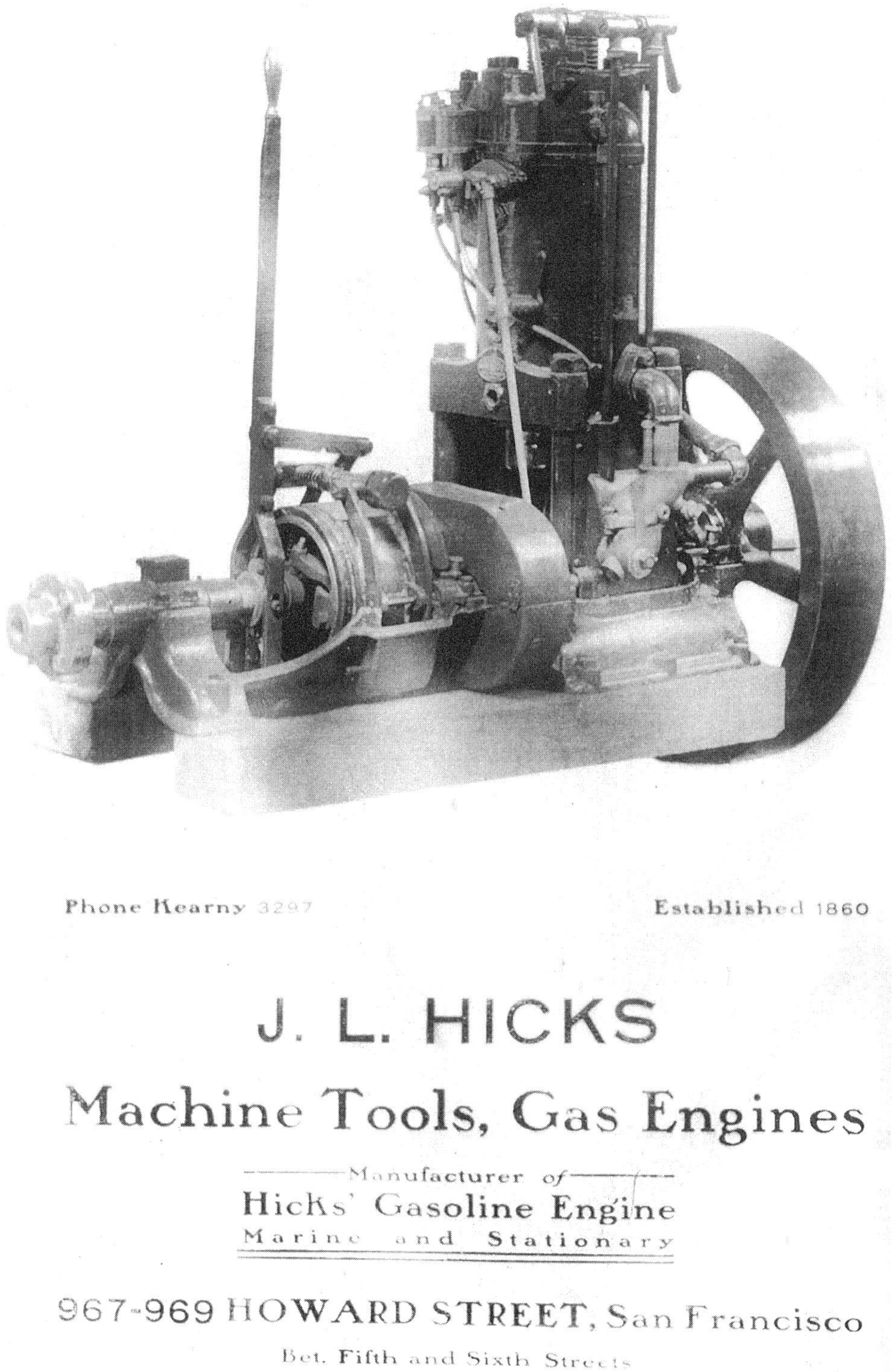

The round cylinder and head of this Hicks indicates it is the model first built by James L. Hicks and later when the company was acquired by Demarest. Subsequent models had a quite different, squared off look. A typically massive, California-style engine, the one-cylinder Hicks weighed more than a thousand pounds. (Courtesy, Greg Hicks)

Gas Engines. Old-timers recall that single-cylinder Hicks engines were available in six-, eight-, and 10-horsepower ratings. Of these, the eight-horsepower was the most popular. The six lacked power for work outside the Bay and was sometimes called the "Delta Hicks" while the 10 horsepower was found to produce excessive vibration.

In 1900, the company name was changed to J. L. H. Machine Tools and Gas Engines and it was sold around 1920 and renamed the Hicks Iron Works. By 1927, when the then-struggling Hicks firm was acquired by the Yuba Manufacturing Company, the sales brochure claimed that "Over 80% of the fishing boats using six- to thirty-five-horsepower engines in California waters are equipped with Hicks engines.... Hicks engines are designed for fishing boat use. They have the valve-in-head mechanism. Manufactured at Benicia, California, where river, bay and ocean fishing is a big business and the needs of fishermen are understood." Yuba expanded the model line to include two-, three-, and four-cylinder models and abandoned the old factory on San Francisco's Howard Street for a plant in Benicia where Yuba also produced gold dredges and crawler tractors. Production ceased in 1932.

Plant of Hicks Engine Company, circa 1918. Fishermen couldn't get enough of the tough, one-cylinder Hicks. Their demand kept the factory busy for years. In this view, Frank Hicks is seen wearing the hat and bow tie. (Courtesy, Greg Hicks)

Two-cylinder Hicks, make-and-break ignition, circa 1927. This is the late model, Yuba-built Hicks. Two-cylinder versions were available in three ratings: 10-14 hp, 12-16 hp, and 16-22 hp. They were rated at 350 to 500 rpm and weighed from 1,830 pounds to 2,400 pounds.

In truth, the West Coast fishing boat market represented a vast opportunity for marine engine builders. Although Hicks had a well-defined portion of the market, there was so much potential that Midwest and East Coast manufacturers established Pacific Coast agencies to represent their products. The Henry C. Hyde Company of San Francisco was just one of these. It handled Speedway engines from Morris Heights, New York, and Peerless engines from Buffalo, New York.

While Hicks, 'Frisco Standard, Imperial, and Union always predominated in fishing boats, a study of magazines of the era shows that West Coast fishermen, particularly those on Puget Sound and the Columbia River were using smaller-horsepower engines shipped crosscountry. Around 1908 or '09, with marine engine sales in the Northwest having attained levels that were simply too high to ignore, representatives from eastern companies began appearing at Puget Sound harbors and elsewhere to learn what was going on. Sometimes, New England *vessels* showed up and began a new life out west. One of these was the Gloucester schooner *Levi C. Woodbury,* which voyaged to Cape Flattery, was fitted with a pair 65-horsepower Automatic engines manufactured in Bridgeport, Connecticut, and by the summer of 1909 was enjoying renewed life as a halibut schooner. Pacific Northwest halibut schooners had first been modified with a gasoline engine beginning in 1903. A decade later, most of these vessels had been redesigned along purely power-boat lines, although steadying sails were carried to control their rolling motion. They were powered by a slow-turning semi-diesel, and later, diesel engine.

Power came to the Columbia River fleet beginning around 1912, and within two years, all the boats had been converted to power. Reviewing the gill netters on the Columbia in 1918, *Power Boating* reported that "The most common makes of engines used are the Standard, Atlas, Union and Imperial. The Regal, Clay and Sterling of eastern make are also used to quite an extent." But other brands were used, too. Following are just a few examples of this trade. In Thorndale, Oregon, boatbuilder Frank Johnson installed three-cylinder, four-cycle Palmers in 11 fishing boats while another Palmer, a single-cylinder, four-cycle was put in a smaller fishing boat in Cascade. A 10-horsepower, two-cylinder, heavy-duty Regal, built in Coldwater, Michigan, was installed in another, smaller boat. So was a little 4 1/2-horsepower Gray. A pair of 13-horsepower Scripps engines were used in two boats that would fish the Columbia and Cowlitz Rivers. A six-horsepower Gray was ordered by a fisherman at Alert Bay, British Columbia.

The switch to power was inexorable. In Washington, on Willapa Bay and in Puget Sound, the seasonal progression of graceful, sail-driven oyster tongers gave way, by 1905, to gasoline-powered boats. These operated pretty much independently of winds and tides and thus shortened greatly the season by stripping the oyster beds clean in comparatively short order. At the same time, wages were good, some boats earning $300 in a month's time, and 200 new jobs were created that year.

At Astoria, the graceful, double-ended gill-net boats began to be built larger as sails gave way to engines. Boats grew from 24 feet to 30 feet, with the average probably 26 feet long. Four- to 15-horsepower engines were typical although the oars were still used to lay out the nets. A *Power Boating* article in 1918 described the matured gill-net boat. "A forecastle

Gill-net boats, Astoria, Oregon, circa 1920. When gill netters began to be converted from sail to power circa 1912, their average length increased from about 18 feet to some 26 feet. Engines ranged from 4 to 15 hp, but oars were still used when pulling off while setting out the nets. The nets themselves can be seen drying on racks on the dock. Some of these boats have what fishermen might have called "fancy fo'c'sles" that were augmented by a tent beneath which a man might find shelter or, on larger boats, even a mattress. At the time this photo was taken, there were some 2,000 boats at Astoria, with gill netters in the great majority. The numbers on the boats may be those of the canneries which, for years, leased boats to fishermen. Eventually, the fishermen owned their own boats, selling their salmon catch to the processors. (Courtesy, Columbia River Maritime Museum)

Tuna boat Alrita, *1944. Built by the Prothero Boat Building Company at Seattle,* Alrita *made her home port in Astoria, Oregon, on the Columbia River. Here we see* Alrita's *engine being lowered into the vessel. The engine is a six-cylinder, 250 hp, Cooper-Bessemer FWDR6. Cooper's sold several such engines to fishermen in the Pacific Northwest that year, each one costing about $15,000. (Courtesy, Knox County Historical Society)*

has been added which, year after year, as the boats have been made bigger, has been enlarged and extended until on the newer boats it might be termed a semicabin.... There is sufficient room under them for a large mattress, so that one man can sleep in comfort while another watches.... The engine is placed a little back of the forecastle, and just back of the engine, extending across the boat, is a 2-feet wide wooden compartment for the fish. On either side of the engine are similar compartments containing on one side a galvanized tank for gasoline and on the other a water bucket, tools, and other small equipment. In some boats, however, the gasoline tank is placed in the bow. Just back of the fish compartment is a space set aside for the net, and a little platform in the stern is left clear for the fisherman to stand on as he throws it out."

The impact of this steady increase in powered fishing boats of all sizes grew to be enormous. The number of motor fishing boats on the Pacific Coast swelled from just a few around 1900 to *30,000* by 1922. On the lower Columbia River, there were then reported to be over 4,000 gill-net and trolling boats in the salmon fishery alone. The result was fish, fish, and more fish — salmon, smelt, halibut, cod and other species in numbers beyond the imagination of those who tracked such things. By 1919, fishing had become the largest single industry on the Pacific Coast, bigger even than timber and mining. In Alaska, it had exceeded gold, copper, and coal. There, the number of canneries jumped from nine to 16 in 1900 and by 1917, there were some 140 canneries in operation. Each was served by tenders powered by a 40- to 50-horsepower engine. While these initially were gasoline engines, the semi-diesel was much in evidence by 1918 and, later, the diesel.

"When fishermen were forced to operate their boats with oars and sails, so much time was lost that it was possible to catch only a small proportion of the number of fish that are now taken by the modern, up-to-date gas-powered fishing boats," reported *Pacific Motor Boat* in May 1922. "It is the motor boat, in fact, that has revolutionized the entire industry, making it more profitable to those operating it and increasing the efficiency and capacity of operation to such an extent that in a few years' time, the aggregate output of the industry has more than trebled."

Regal engine, Astoria, Oregon, 1918. These heavy-duty, four-cylinder engines, built in Coldwater, Michigan, were well regarded by the salmon trollers of Astoria who installed them in boats that averaged about 26 to 33 feet in length. By 1918, trollers, which were equipped with four or five poles made of young saplings, were becoming ever more popular, for they did away with the expensive nets of the graceful gill netters. Such nets cost $300 apiece and most fishermen needed a pair of them.

The magazine listed the type and size of the fleets as they then existed. There were salmon fleets in Sacramento and at Monterey; tuna and sardine fleets at San Pedro and San Diego; purse seiners on Puget Sound, gill netters on the Fraser River; a fleet of halibut schooners operating in Southeast Alaska and Puget Sound; the crab fleet at San Francisco. There was a big fleet of trollers at Cape Flattery and, biggest single fleet of all, the several thousand sturdy Bristol Bay boats. There were times, during the great conversion to power that builders of halibut schooners and gill netters couldn't keep pace with demand.

It is easy to forget, in the very last years of the 20th century, just what the coming of power meant to the commercial fisherman at the century's beginning. Technologically and culturally, the world of 1999 has little in common with the world of 1899. Although fishing methods developed over time, and although boats were refined over time, the late-19th-century fisherman still did his work under the power of oar and sail. The impact of steam, while felt in some fisheries, left the great majority untouched. But the first gasoline engines changed everything overnight. It was only where legislation prohibited the use of power, such as in Chesapeake Bay oystering, on the oyster beds of the Mississippi, and the Alaskan sealing fisheries, that work continued to be performed under sail.

But otherwise, engines were in use throughout the country by the very early 1900s, were pervasive by 1910, and dominated the industry a decade later. For the fisherman, installing an engine was a matter of economics. Adopt power or risk the results that accompanied becoming less productive and, ultimately, non-competitive. On the bigger vessels, which generated significant revenues, the engine was paid for by the vessel's owner and was soon taken for granted by crews. But those fishermen had little to do with the machine itself, which was run and maintained by an engineer.

It was the man who fished with one or two of his fellows, and especially the lone fisherman, to whom the engine meant most in a personal and enduring fashion. In the early days, on the Atlantic coast from Newfoundland to New England, and in dozens of other spots, the engine was almost invariably a one-cylinder, two-cycle machine, usually with a make-and-break ignition system that would work reliably in an open boat. Such an engine would forever be referred to as a "one-lunger," after the distinct intake and exhaust sound — which could be likened to one's own breathing — emitted by its single cylinder. Often, the engines were also referred to as "make-and-breaks," a reference to their low-tension ignition devices or "igniters" that reliably took the place of a spark plug. These were engines that, once understood, would last a lifetime, but their acquisition was not necessarily a simple matter.

"A four-horsepower Acadia would cost from $350 to $400 in Canada in those early days," said fisherman John Budgell of Northwest River in Labrador. "That was a considerable sum and a fisherman would have to have two or three good summers to be able to afford one. Of course, the price paid went up with the horsepower. The price of a motor would be just about what the fisherman needed for his winter's stash of grub and, if he did not have a trapping place for the winter months, he very often had to do without his motor."

Once they got such an engine, the Newfoundlanders, in particular, kept them. Stories of the methods used by self-sufficient fishermen to keep their one-lungers going have long since passed into legend. In part, these heroic measures became necessary simply because the engines sometimes outlived their manufacturers, and spare parts were not always attainable. Budgell remembered an Eskimo fisherman whose engine's cylinder head was bolted to the cast-iron cylinder. Normally, a removable head was a desirable feature but on this engine, the head bolts had been stripped and the engine had lost compression.

"He searched around in some shed and came up with a pair of anklecuffs," remembered Budgell. "These had a long chain between the cuffs and this he used to wrap around the rear end of the engine and the head, and then he twisted the whole thing tight with a large nail or bolt and away he went for a few more seasons."

Budgell knew another fisherman whose eight-horsepower Acadia finally lost all its compression out along the main bearing in the middle of central Labrador's Lake Melville. "He lifted the cylinder off the base," said Budgell, "cut a piece off his leather belt, fitted it in the bearing and was on his way. In fact, it worked so well he left it there for the rest of the summer."

When engines like these were finally, as the Canadian fishermen used to say, "past their labor," they were replaced by a small diesel or an outboard. In Newfoundland and Labrador, these replacements were still going on, incredibly, in the 1950s and 1960s. And the new machines were generally regarded, in comparison to the old one-lungers, as being fast, noisy, smelly, and a nuisance in ways the old engines never were. Of course, they were not nearly so easy when it came to improvising a repair.

And so, in Newfoundland, where fishermen took to the two-cycle engines in a special way, in the remote little outports now mostly abandoned, and in the bigger towns, and St. John's, too, the sons and grandsons of those old fishermen have not forgotten the first engines, or what they meant. These were stalwart, cast iron machines from Acadia and the Lunenburg Foundry in Nova Scotia, and from Palmer and Mianus in the United States. There were others, as well, too numerous to mention, for the market never really seemed totally fulfilled. One day, in August 1992, when composer and writer Ray Johnson sat down in Job's Cove, Newfoundland, to write a poem about how those old-timers built their boats and what they thought of their engines, he said it simply. To the fisherman of those long ago days, he wrote, the make-and-break marine engine was "a gift from heaven."

GM diesel, Detroit Auto Show, 1949. GM's contribution to the war effort was widely recognized and its diesel engines were the centerpiece of the effort. The engines soon became a dominant force in the postwar market for marine and railroad power. (Copyright 1999 GM Corp. Used with Permission of GM Media Archives)

Chapter 4

General Motors

"At the present time, my opinion of the Diesel engine is not fit to put into print."

—Charles F. Kettering, July 24, 1928

"With few exceptions the engines were large, heavy, and slow running, and so found their greatest application in power stations, pumping, and marine use. They weighed 200 or 300 pounds per horsepower, and this, indeed, was the heart of the problem — to build a powerful, fast-running diesel of relatively small size."

—Alfred P. Sloan, Jr., *My Years With General Motors,* 1963

"Well, I'd say that he was one of the very greatest men."

—Cyrus R. Osborn, Vice President, General Motors, regarding Charles F. Kettering

---

One gentle spring night in May 1945, a business executive named George Codrington gave a speech to the American branch of the Newcomen Society of England. The Society's membership was dedicated to "continue the struggle towards a nobler Civilization through wider knowledge and understanding of the hopes, ambitions, and deeds of leaders in the past who have upheld Civilization's material progress." At the time of his speech, Codrington was the General Motors vice president responsible for the corporation's Diesel Engine Division. GM's various diesels had become a hugely successful part of the war effort, and Codrington himself had played a role in what was certainly the longest and most extensive diesel development program in the United States. He crafted his

Alexander Winton (1860 - 1932). By the time he died in Lakewood, Ohio, Winton was widely recognized as one of America's leading automotive pioneers and as one who did much to develop a practical marine diesel engine. His Cleveland-based Winton Engine Company was acquired by GM and played an important role in the latter company's diesel-development program.

speech that night to pay tribute to the "inventive achievements of two great leaders in the field of Engineering — two men whose shadows and influence can be detected today."

One of the men was Rudolf Diesel, whose work, Codrington recognized, had influenced an entire generation of engine men. The other individual was a man whom Codrington knew personally, for he had gone to work for him in 1917. He had been born in Scotland in 1860 and he went, as Codrington said, "from an unknown Scottish lad struggling for a living in New York City to a world-famous inventor and bicycle builder, as well as pioneer builder of motor cars and marine and industrial power plants." This man was Alexander Winton.

In 1879, Winton arrived in New York City, and there, what Codrington called "his mechanical cleverness" soon won him a job in a marine-engine shop. He stayed in New York long enough to attain a mastery of steam engines and then shipped out as a licensed engineer on ocean-going vessels. It was a period in Winton's life that established a profound love of the sea and it would lead, in later years, to the creation of a remarkable marine engine company of his own. But first, there were bicycles. With his seagoing days behind him, Winton moved inland to Cleveland in the late 1880s and took a job at the Brown Hoisting and Engineering Corporation. He left this job to start the Winton Bicycle Company in 1891. The cycle and the timing were both excellent. Besides establishing a solid base of sales in the United States, Winton exported his bicycles, too, and he made some serious money. This he used, beginning in 1893, to fund experiments with a horseless carriage. Thus was planted the seed for what became, in 1897, the Winton Motor Carriage Company. In April 1898, Winton sold his first car, a $1,000, single-cylinder machine, to a mechanical engineer from Port Carbon, Pennsylvania, named Robert Allison. Later, Allison would recall that he had visited "various places where horseless carriages were said to be in process of manufacture but he found that inventors were still inventing. Several inventors were eager to sell him stock but not one of them would agree to deliver a machine guaranteed to run. Naturally he was pleased to find Mr. Winton prepared for actual business."

Allison's was the first of 21 such vehicles sold by Winton that year, making him the first man to commercially market an automobile in any volume. His customers were wealthy men whose professions Winton was careful to note: a piano manufacturer from St. Louis, a locomotive manufacturer from New

Jersey, coal mine owners from Pennsylvania, "capitalists," and, from nearby Warren, Ohio, a maker of electrical equipment named James Ward Packard. Two years later, Packard would start a car company of his own. As for Robert Allison, he later bought several more Wintons, including a two-cylinder, a four, and a six, introduced by Winton in 1906. These were all grand vehicles, for Winton had positioned himself at the topmost echelon of the industry. It was a strategy that worked well until the financial decline and market changes that followed World War I. Stubborn and demanding, Winton liquidated his car company in 1924 rather than introduce less expensive models.

The end of automobile production gave Alexander Winton more time to pursue what George Codrington said was Winton's "first love all his life, the sea. This is not strange, in view of the Scotch blood that flowed in his veins, and his early experiences as a boy working in the shipyards along the Clyde." But his automobiles gave Winton the financial resources, estimated to be $5 million, to indulge his interest in the sea as a yachtsman. He had spent some of this money on a steam yacht, the *La Belle.* His problem with the *La Belle* was that he had to let the professional captain know well ahead of time just when to have steam up. It was an inconvenience.

In 1911, Winton decided to have a big, gasoline-powered motor yacht constructed. Cox & Stevens designed a new *La Belle* that was 133 feet long and Winton, like his own early automobile customers, began visiting marine engine companies to examine their products and make a selection. None were satisfactory, either because of noise or vibration. Instead, Winton designed a six-cylinder engine with a nine-inch bore and 12¾-inch stroke that was rated at 150 horsepower. Within four months, the patterns for the engine had been made, castings had been poured and machined, and three engines were built and installed in the *La Belle.* They proved to operate so quietly, so dependably, and so cleanly that Cox & Stevens, among others, suggested that Winton should manufacture them. The Winton Gas Engine and Manufacturing Company was organized in Cleveland in February 1912, and a new 100- by 300-foot factory was built on Cleveland's West 106th Street near the Winton automobile plant. Operations got off to a slow start. There were those in the marine engine business who did not relish having Alexander Winton as a competitor. No sales were made the first year and the business in general struggled for several years before it became self-sustaining.

Winton's engines followed the general pattern of the *La Belle*'s. These were heavy-duty, six-cylinder models that included the 80-horsepower W6, the 125-horsepower W5, and the Model 11. The latter's bore and stroke were increased in dimension over the *La Belle*'s engines and measured 9½ inches by 14 inches to provide 200 horsepower at up to 475 rpm. These Winton gasoline engines were mighty L-head machines that ranged in weight from the 5,400-pound W6 to the 19,000-pound Model 11, and they remained in production with no major design change until the early 1930s. During World War I, the eight-cylinder W29 was introduced, a 300-horsepower model that was sold in some quantity to the U.S. Navy for use in minelayers and to the Russian Navy.

Exactly when Winton became fascinated by the possibility of diesel engines has gone unrecorded but it must have preceded or else coincided with the building of the *La Belle*'s gasoline engines. "From that time until the day he died," according to George Codrington, "Alexander Winton's life was devoted to the development of Rudolf Diesel's brainchild." The first Winton diesel, an engine that Winton and his colleagues believed was the "first American-designed internal combustion engine employing the Diesel principle of ignition and compression," appeared in the fall of 1913. This six-cylinder engine was installed in the Winton automobile factory to generate electricity. A production version, the 175-horsepower W-16 was then introduced. Like other diesels of its era, it was a cast iron machine with individual cylinders bolted to an iron upper crankcase. A three-stage air compressor was driven off the front of the crankshaft.

The six was followed by a 300-horsepower V-12 in 1915, a pair of which promptly took the place of the gasoline engines in the *La Belle.* These engines, which Winton designated as the W30, must almost certainly have been the first V-12 diesels ever made. In 1916, the W24 engine was unveiled, a 225-horsepower, six-cylinder marine model with a 12 15/16-inch stroke and 18-inch bore. It was followed by an eight-cylinder, 500-horsepower version, the W40.

Thus was launched a line of engines, and a company, that was destined to play a central role in American diesel-engine development. George Codrington, who had met Winton in 1913, joined the company four years later as its factory superintendent. He rose quickly, being named general manager in 1919, vice president in 1925, and president in 1928. That was the year that Alexander Winton sold the company to an investment group, although he remained as chairman. Looking back on his initial challenge at the company, Codrington said it was to increase the customer base, which he did, in part, by increasing the emphasis on engine sales to the U.S. and allied navies. He also pushed sales to the commercial marine market and Wintons were installed in a variety of big, auxiliary freight schooners ranging from 230 to 255 feet in length. He was quick to capitalize on Winton's successful efforts at diesel-electric drive, begun with the installation in the 162-foot, Herrreshoff-designed schooner yacht, *Elfay* in 1919. This big yacht, owned by R. A. Alger of Detroit, was going to be used for a world cruise and the vessel was taken to the George Lawley yard at Neponset (Boston) to have her steam engine removed and a diesel installed. It was a Model 59, six-cylinder, 125-horsepower Winton driving a 75-kw generator that operated a 90-horsepower electric motor connected to the boat's propeller shaft. An ex-Navy engineer named J. S. Walters of Spokane, Washington, who had served aboard the submarine N-2, was put in charge of the vessel's machinery, which also included a 25-horsepower generating set. *Elfay* was said to have been the first diesel-electric yacht, and Winton soon found many more such applications.

By the mid-'20s, Winton was recognized as a premier supplier of diesel engines for yachts of 120 to some 190 feet, as well as workboats. The company became a regular advertiser in boating publications, often showing a big photograph of the latest vessel to be equipped with a diesel electric or direct drive installation. It was a kind of golden age in yachting, and significant wealth was lavished on elegant boats. Of the vessels classified as "yachts" by Lloyd's during the '20s, it was estimated that Winton powered some 57 percent of those equipped with diesels, and 61 percent had Winton gasoline engines. Approximately 500 of the latter were built before operating costs moved the market to diesels.

Frequently, Winton diesels replaced steam engines. Such was the case with the 600-ton, Lawley-built schooner *Alcoyne.* A pair of 225-horsepower Wintons driving a Westinghouse generator replaced the original steam installation. The other key development upon which Winton embarked after the close of World War I was the development of an airless-injection fuel system. Like everyone else, Winton engineers wanted to replace air injection with its heavy air compressor and all the difficulties encountered in holding high- pressure air in tubing and various fittings. An airless-injection Winton was introduced in 1928. It used a three-cylinder, plunger-type pump to pump fuel into the manifold (common rail) and, from there, it was injected into the cylinders by a camshaft-operated injector valve. A pair of these 700 rpm engines was promptly ordered for installation in a 94-foot Matthis cruiser. They were handsome machines, fully enclosed, with a one-piece block cast of chrome-nickel iron that carried inserted cylinder liners.

Airless injection was a big step for Winton. Engineers had already determined that, added to all its other deficiencies, the air-injected diesel was some eight percent less fuel efficient than a common rail injection engine. But it was also recognized that common-rail injection was known for "dribbling," the minute leakage of fuel after the injection was supposed to have ended. Carbon buildup and poor combustion resulted. What's more, by the early '20s, the entire matter of how fuel was injected into the cylinder of a diesel engine had taken on significance that went beyond simply doing away with the air compressor and high-pressure air lines. Both in Europe and the United States, the related issue became one of reducing the size and weight of the engines while increasing rpm so that horsepower to weight ratios would be dramatically improved. The goal was creation of an "automotive" diesel suitable for a variety of applications.

These were the issues central to the advancement of the diesel engine, and they were recognized as being so far-reaching and difficult that they actually brought together men from competing companies for discussion. In 1923, at Pennsylvania State College (now Pennsylvania State University), engineering dean R. I. Sackett began investigating the distribution of fuel spray inside combustion chambers. Later, in the spring of 1927, the college hosted the first of many meetings — the Oil Engine Power Conference — attended by diesel engineers and executives to discuss commonly faced problems.

Even as the Penn State meetings were initiated, ways of building a high-speed, lightweight diesel were being discussed elsewhere in the country. In March 1927, the Society of Automotive Engineers had a meeting in Indianapolis to discuss the high-speed diesel engine. Principal speakers included Clessie Cummins whose company had learned the hard way the challenge of developing a reliable, lightweight diesel engine, and W. W. Schettler of the Foos Gas Engine Company, which had recently introduced a new, compact diesel model.

Methods of improving the diesel engine were now being investigated by companies as diverse as M.A.N. in Germany, Winton in Ohio, Cummins in Indiana, and by Otis Treiber in New Jersey. Treiber, in fact, had left his post as engineer at Bessemer in Grove City, Pennsylvania, to establish his own company to build a lighter-weight, high-speed diesel. The production Treibers were beautiful engines, fully enclosed, overhead cam models that, according to Treiber himself, "occupy about one-third the space and scale about one-sixth the weight of earlier type diesels."

Keeping tabs on all of these developments was the presiding genius of the American automotive industry. Just about everybody in General Motors called him "Boss" except for the company chairman. Alfred P. Sloan, Jr. called him "Ket." He was Charles Kettering.

One day in 1928, Sloan paid a visit to Kettering, which he recounted in his book, *My Years With General Motors.* "If my memory serves me correctly," wrote Sloan, "I remember dropping in one day at the Research Laboratories in Detroit and saying to Mr. Kettering: 'Ket, why is it, recognizing the high efficiency of the diesel cycle, that it has never been more widely used?' In his characteristic way he said the reason was that the engine would not run in the way that engineers wanted it to run. I then said to him: 'Very well — we are now in the diesel engine business. You tell us how the engine should run and I will see that available manufacturing facilities are provided to capitalize the program.'"

Like so many highly placed automotive executives, Charles Kettering had a motor yacht. In 1928, the boat he owned was one powered by a Cooper-Bessemer, four-cycle diesel. It had a typical common rail fuel injection system that, Kettering found, would inject the desired amount of fuel at cruising speed while not feeding fuel to all cylinders in equal quantity at other speeds. Kettering did some serious tinkering with this Cooper-Bessemer that improved operation somewhat but not nearly enough to satisfy him. He wrote that "I rebuilt my Bessemer injection mechanism and finally got the engine to run in control in a fairly satisfactory way.... The great problem is still that of injection control and if you are going to try to get your engines to work over a 75% speed range, it is quite a problem."

Kettering perceived that he'd done about all he could with his current boat and determined that a new yacht, with new engines, would enable him to further develop his understanding of what was needed to make the diesel engine more satisfactory. The result was a boat which, like the previous vessel, was named *Olive K* after Boss's wife. The *Olive K* was a 170-footer designed by Cox & Stevens. Those who knew Kettering liked to joke that this boat, which was lavishly appointed and had a phonograph that piped music to the staterooms, was really an engine room around which a yacht was constructed. Alfred Sloan wrote that "as anyone who knew him might have guessed, when on his yacht he was more often tinkering in the engine room than relaxing on deck."

Charles Kettering was then famous to many Americans as the man who had invented the electric starter for automobiles, quick-drying lacquer paint, safety glass, the high-compression engine, and no-knock gasoline. During his lifetime, he received 140 patents and amassed a vast fortune, although his colleagues often picked up the lunch tab because the Boss had forgotten to bring pocket money to work. Born on a farm near Loudonville, Ohio, in August 1876, Kettering was studying classical languages at Wooster College when he switched to Ohio State to study engineering. At age 28 in 1904, after a two-year interruption caused by an eye disorder from over-study, Kettering graduated with degrees in mechanical and electrical engineering. This training was soon put to use at the National Cash Register Company where Kettering did, for the first time, something that more experienced men said couldn't be done: He invented an electric-motor-driven cash register. In 1909, he founded the Dayton Engineering Laboratories (Delco) with his associate, Edwin A. Deeds, and the men created an ignition system with a storage battery rather than a magneto. Delco's next step was the electric starter.

Given Kettering's background and his already renowned individualism, the yachting press was not entirely surprised by some of the features in the second *Olive K*. "Perhaps never in the history of yachting has there been such an unusual craft as *Olive K*," reported *Motor Boating* in January 1930. "In the opinion of its builders, the Defoe Boat & Motor Works [Bay City, Michigan], this yacht is expected to push the industry ahead ten years. The boat had stabilizers that could be deployed to reduce roll, carried a 24-foot 'sedan,' a 22-foot 'speedster coupe,' plus a pair of more normal 21-foot boats, and its diesel engines were electrically synchronized to reduce vibration." Of course, it was the engines themselves that were Kettering's chief objects of interest. They were Wintons.

Charles Kettering (1876 – 1958) in the General Motors Research Laboratory, circa 1932. An inspirational leader and brilliant engineer, Kettering — known as Boss Ket to his colleagues — was the driving force behind GM's decade-long program to develop the high-speed two-cycle diesel. (Copyright 1999 GM Corp. Used with Permission of GM Media Archives)

Although General Motors had made some quite modest diesel investigations in 1920 and 1921, it was in 1928 that the corporation's Research Laboratories began, in a small way, what would become a major development effort. Kettering's friend, Carl Fisher, the multimillionaire yachting enthusiast who had developed Miami Beach, was predicting "an enormous future for Diesel motors in both marine work, railroad work and in the truck business." Kettering was also hearing from his friend Henry Joy. Packard's president urged him to consider for the new *Olive K* the diesel engines developed by Otis Treiber. Kettering did so but found he had significant reservation about the injection mechanism. He was more interested in the work then being done by Cummins, but he seemed more interested still in a project then underway at Winton. There, chief engineer Carl DeWitt Salisbury was at work on a type of fuel injection that was based on an individual injector for each cylinder, a self-contained "unit injector."

Although Kettering had initially decided against buying Wintons for the *Olive K*, Winton's sales manager persuaded George Codrington to let Kettering see what the company was then calling its C.D.S. (Carl DeWitt Salisbury) Airless Injection. Kettering said he'd order Wintons if the company would equip them with these unit injectors rather than common rail injection. The Winton response was that the injectors, which Salisbury had been working on since at least 1926, were not yet ready. Kettering's response was that he wanted them anyway, that if worse came to worst, they could revert to the common rail system. That is how the *Olive K* was equipped with a pair of six-cylinder Wintons, the order placed November 16, 1928.

"I went to Winton Engine simply because they would put in a different type injector," Kettering would say later. Each of these engines had a 14-inch bore and 16-inch stroke. They developed 500 horsepower at 350 rpm and weighed 48,000 pounds apiece, a power-to-weight ratio of 96:1, this at a time when Packard had introduced an aircraft diesel. Winton was right to have reservations about supplying Kettering with unit injector-equipped engines. The injectors weighed 75 pounds each and, as Salisbury and others at Winton had feared, didn't work. "It was no good — broke right away," remembered Kettering. His old colleague, Edwin Deeds, who had also bought Winton engines for his own new boat, suffered the same problems.

By this time, the GM Research Laboratories had been at work on its own unit injectors for some two years. This development was conducted in conjunction with a single-cylinder, two-cycle diesel test engine that, with an eight-inch bore and 10-inch stroke was known in the Lab as "Big Bertha." As it happened, there was a series of single-cylinder test engines built, both two- and four-cycle, each an improvement over its predecessor. Much of this development work was conducted by Frederick Glen Shoemaker, who had joined the Lab in October 1928 at age 39, his previous job having been at the Franklin automobile company. A machinist and mechanic, Shoemaker had been fascinated by steam engines as a boy and then graduated from the University of Illinois at Urbana with an engineering degree. Working closely with him was Bjarne Thulin, a Norwegian-born diesel engineer. Both men were well aware, as was Kettering, of experiments at Studebaker with fuel-injected gasoline engines, and they also knew about Salisbury's work at Winton.

In his personal logbook in October 1928, Shoemaker had noted: "Engineers of the Winton Engine Company presented drawings of a new fuel valve which incorporates a pump in each nozzle assembly operated by a variable stroke valve gear. The pressure from the pump is applied to a very small volume of oil confined in the rigid body of the nozzle, thus eliminating all tubes and internal effects. The spray nozzle consists of a light needle seating in the tip of the spray nozzle and operating as a differential piston by the fuel pressure."

"Big Bertha" experimental engine, GM Research Lab, circa 1928. A series of single-cylinder test engines with an 8" bore and 10" stroke formed the basis of GM's two-cycle diesel-development program. (Courtesy, GMI Alumni Historical Foundation)

As future results would demonstrate, Salisbury's fuel injector would only work on a slow-turning engine and, initially, not even on that. Now, the Research Lab stepped up its work on a unit fuel injector with the idea of replacing those in the *Olive K*'s Wintons. Among other things, the injectors, which developed some 20,000 pounds of fuel pressure, tended to break or bind. "The successful modification built by Research had at least this high a pressure minimum and could, according to Kettering, go up [to] 35,000," reported Don Borlund in an unpublished manuscript on the Lab's diesel program. These pressures were far higher than the 3,000 to 7,000 p.s.i. of a common rail system.

Lab records showed that "the construction of the sample unit injector for Kettering's boat continued in the Research Lab and the injector was finished in September." On Sept. 11, 1930, Shoemaker noted that Kettering asked that the injector for the boat engine be "polished and chromium plated to demonstrate

Charles F. Kettering aboard the Olive K. *Boss Ket replaced the solid-injection system of the vessel's Winton diesels with unit injectors. (Courtesy, GMI Alumni Historical Foundation)*

what a good injector should look like." After various difficulties and a considerable amount of overtime, the unit injector was finally finished about 10:30 a.m. on September 19, 1930. It was taken to Kettering's yacht at the Detroit Yacht Club and installed that afternoon."

The *Olive K* immediately set out on an 18-hour cruise to Bay City, Michigan, during which the injectors required no attention. What's more, the boat ran into heavy weather and her skipper, Captain V. J. Shaughnessy, reported that the engines "were the first marine diesels he knew of that could be throttled down sufficiently to allow the ship to just hold her position and ride out the storm (about 50 turns of the shaft per minute). The accurate metering of fuel plus good fuel atomization of the unit injectors made this possible."

Kettering's old colleague, Edwin Deeds, was not so fortunate. Although Kettering had tried to have Winton build enough new unit injectors for both himself and Deeds, this proved impractical and Deeds' boat was converted to common rail injection. But Kettering kept Deeds informed about ongoing developments and, in November 1930, reported that "In this new design we have made many changes.... we have reduced the amount of oil under compression to about 60% of the old valve... It is possible to stop these engines and then bring them back to any given speed and the temperatures will come back on all cylinders to what they were before. This, I believe, is the first time in the history of diesel engines, where such a control has even been accomplished.

"All of this new stuff is no different than any other new thing and it has taken a lot of mental and moral effort to keep the thing from being discarded and the things put back in the old-fashioned way. Now since we have finally gotten this job working satisfactorily and it has come out as per our original design, the yacht architects are naturally becoming very interested. Up to the present time they have had just about as much enthusiasm as a piece of cold liver..."

Even then, the injectors still required further development. Shoemaker reported that on a cruise in the Caribbean, Boss spent a continuous 24 hours in the *Olive K*'s engine room working on the unit injectors rather than call for a tow. He actually did get the big diesels going again, too. According to Shoemaker, those at Winton might well have taken the opportunity to revert to common rail injection, but Kettering brought the fuel injectors back to the Lab and went to work.

"As a matter of fact," Shoemaker said, "it was found that the injectors had been made using stainless steel, which is notoriously bad bearing material. This, in connection with some of the design features of the injector, caused the plungers to stick and the injectors to fail. By using different materials and moving the pumping unit to the upper end of the injector assembly away from the heat and making slight modifications in the spray valve assembly, it was possible to rework these original injectors." Those Winton engines with their GM-developed unit fuel injectors were still operating in the *Olive K* in the late 1950s by which time the yacht had been converted into a New York harbor pilot vessel.

At the time that Kettering first began talking with Winton about engines for the *Olive K*, General Motors had actively been engaged in the hunt for a diesel engine manufacturer to add to the corporation. Shortly before Christmas 1928, Kettering wrote to GM's chairman: "My dear Alfred, I have been making a study of the engine situation from the standpoint of an independent company which would manufacture engines of various types. . . . To purchase or make some arrangements with the Cummins Engine Company at Columbus, Indiana, who is now making what we consider the best small Diesel engine on the market. They have a fairly good business now and I think, are making a little money. . . . If you wanted to extend this plan further, then my next recommendation would be to acquire the Treiber Engine Company at Camden, N.J. This is a small concern and the major part of its assets consists of Treiber himself. . . . He is a very unusual type of fellow and would make a successful business manager for this entire project. If you still felt like making this a large operation, then my next suggestion would be to purchase the Winton Engine Company of Cleveland."

It was a typically ambitious plan. The Cummins issue, however, seems not to have been considered very seriously. At the time, the Columbus-based company was entering a period when, if hoped-for sales to railroads had fallen through, the marine engine side of the business was doing well. A $200,000 GM offer was made to Treiber who accepted only to cancel shortly thereafter because of a license contract he'd signed with Consolidated Shipbuilding. On October 21, 1929, GM vice president John Pratt visited Winton to inspect the plant and immediately recommended that Winton would be a good "vehicle for capitalizing the developments of our research organization along engine lines and will assist materially in keeping us abreast of Diesel engine developments."

Then came the stock market crash. Its dramatic impact on GM shares led the corporation to withdraw from the proposed deal. Within a week of the crash, word came to suspend raises or hiring and to cancel plans to purchase a M.A.N. diesel for study. Major George E. A. Hallett, who, as a young Army officer, had been assigned to work with pioneer aviator Glenn Curtiss and later was hired by Kettering and eventually placed in charge of the GM Lab, battled hard to get projects restored. Publicity attendant to Packard's diesel airplane engine apparently helped to convince the highest management of the need to continue diesel

WINTON-DIESEL

Powers

50-Foot Wheeler Stock Cruiser

Most Delightful, Practical Boat of Its Size Ever Offered

THE new 50-foot Wheeler Playmate Cruiser, *the first stock boat powered with modern Diesel engines,* marks another and most important Wheeler contribution to the advancement of pleasure afloat. This safe, efficient, seagoing cruiser will be powered with a New-type, high-speed, light-weight, 100 h. p. Winton-Diesel Engine, adding low cost operation and increased cruising range to the unsurpassed reliability and sturdiness of the Wheeler Playmate. Designed especially for Diesel power, each boat is built individually and planned to provide a beautiful and comfortable cruising home—complete and carefully planned in every detail. The New-type Winton-Diesel, developed expressly for craft of this size and type, permits the ideal combination of a smartly designed stock boat and modern, economical, Diesel power. Complete details upon request.

New-type 100 H. P. Winton-Diesel

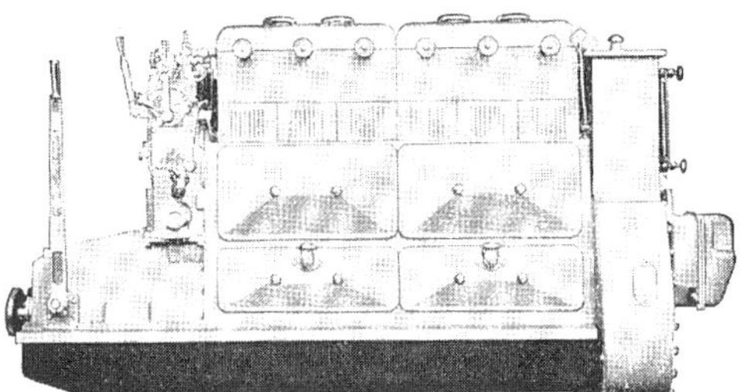

Selection of this New-type Diesel as standard in the 50-foot Wheeler Playmate for 1932 marks the introduction of a high-class Diesel-powered stock cruiser. Modern Diesel power, provided by this highly developed, compact, smooth-performing new engine, is destined to make this cruiser a most popular stock boat.

WINTON ENGINE CORPORATION

CLEVELAND, OHIO, U. S. A.

Winton was part of GM at the time this ad appeared in 1932.

research at the Lab. Hallett was permitted to purchase a six-cylinder Mercedes-Benz, and a two-cycle Junkers for testing.

At this time, too, William S. Knudsen, then head of Chevrolet, was very interested in the potential of a three-cylinder, two-cycle diesel truck engine that Kettering was having sketched out. A go-ahead was given for building a prototype. The engine's fuel system was based on a modified Bosch fuel pump with an individual plunger pump feeding an injector valve at each cylinder. This engine was beset by excessive vibration, inadequate torque, smoking and poor fuel consumption. It was obviously unsuccessful from a commercial standpoint but, as a laboratory experiment, it provided invaluable lessons. It also suggested that Kettering's instinct about the two-cycle diesel's promise for offering exceptional power-to-weight ratios was correct. That 50-horsepower engine weighed 835 pounds, a 17:1 ratio. The lab's 45-horsepower M.A.N and 92-horsepower Buda had ratios of about 33:1 and its six-cylinder Mercedes-Benz, about 22.5:1 thanks to use of aluminum wherever possible.

Finally, in May 1930, George Codrington received an official offer letter and the purchase was completed on June 20, 1930. Winton Engine Company was renamed Winton Engine Corporation, a wholly owned subsidiary of General Motors. It operated as such until 1934 when it became the Winton Division of General Motors and then, in January 1938, the Cleveland Diesel Engine Division of General Motors. Years later, when Boss Ket was asked why it had been Winton that was purchased by GM, he replied, "Because they were closest here." By then, he had his reasons for such a cryptic reply.

Winton wasn't GM's only acquisition in 1930. Electro-Motive, a struggling maker of rail cars powered by Winton gasoline engines, had plans for a lightweight diesel engine version but didn't have the capital to develop it. The company's president, Harold Hamilton, estimated it would cost $5 million to develop a diesel with the power-to-weight ratio needed for railroad work.

"We didn't have that kind of money available," he remembered in a 1957 interview. "Neither did Winton. And even if they had, the money itself would not have been enough. Winton didn't have the technological men and equipment required for development of the engine we needed. General Motors did."

Electro-Motive entered the GM fold in November 1930. Now, GM had prepared the way for entirely new diesel engines that would combine the know-how gained by GM research with the practical, diesel engine building skills of Winton and the railroad-oriented expertise of Electro-Motive. Almost at once, Kettering began thinking of building another new yacht, this one with two-cycle, unit-injected diesels. But he soon determined that the necessary technology remained to be developed.

Winton was, however, then at work on a concept aimed at reducing a big diesel's weight, if not increasing its power. It had developed a crankcase built up of welded steel rather than cast iron. Winton contracted with a specialty firm named Lukenweld to construct the welded structure, and the first engine built in this manner was still operating reliably 17 years later. This was an innovation that, when proven by testing, would have enormous impact on power-to-weight ratios, for it lowered the weight of the crankcase by some 20 percent. Meanwhile, work at GM continued on Big Bertha and, in 1931, the four-cycle model being tested was converted to two-cycle operation, and tests of a blower for such a machine were also ongoing. The lab was working intensely all throughout the 1930–'31 period, even as the Depression forced a rather substantial cut in manpower. There was even brief talk of cutting the diesel program. Some of those let go were hired by Winton. Both the Lab and Winton continued unit injector development, each running into one problem after another. Brief notes give insights into the problems: "did not follow the cam at 825 rpm," "passed too much oil when supplied with fuel at 20–25 pounds pressure," "new needle guides split," 'no hope of producing suitable parts for any unit injectors until this new precision equipment is set up and the personnel trained to do the required work." Of course, these problems related only to the unit injectors. Before multicylinder versions of Big Bertha could ever be considered viable, there were issues relating to everything from noise and vibration of gears and side panels to leaky packing on the cylinder liners, head gasket problems, crankcase ventilation ills, and a long list of other challenges. One solution after another was investigated until an answer was found. It was Kettering's way.

"One of the characteristics of the Boss that showed up in our discussions of engineering problems," said Glen Shoemaker, "was his skepticism about the validity of using purely technical analysis based on mathematics and such other devices to arrive at a final conclusion. In every case, he insisted that the thing be put to the test to see if the engine gave you the same answer as formulas did. And this in large measure accounts for the cut and try laboratory program that was followed all through the diesel program."

Although Winton and the Power Plant section of the GM Research Laboratories each had rather specific assignments, things did not always go smoothly. The Winton group was viewed by those at the GM Lab as being overly conservative. Those at the lab, with its Kettering-inspired focus on two-cycle development, wanted to break new ground. After all, they went to work each day for a man who might drop by and leave them with a statement like "Failures, repeated failures are the finger posts on the road to achievement…one fails forward toward success." Or, "The Wright brothers flew directly through the smokescreen of impossibility." But underlying the knotty development problems at Winton in Cleveland and the GM Lab in Detroit, as Borlund recorded, was the potentially nasty "old friction between the proponents of old proven ideas and methods and the enthusiastic backers of the radically new, potentially better way."

The whole point of Big Bertha was to use the experience gained at the Lab with a single-cylinder engine and then adapt it to a multicylinder unit. Yet, when the first eight-cylinder two-cycle diesels were built at Winton in 1933, "Mr. Kettering said we have thrown away all our experience on the 8 x 10 single and built the 8 in-line jobs all different and no d—n good!" reported Major Hallett.

Boss was particularly unhappy regarding details that had been proven out at the lab — such as the location of governor shaft control arms and the elimination of clumsy rocker shaft supports — but had then apparently been ignored. One thing, it was acknowledged, that was sure to upset Boss was to have a design modified after he had passed on it. So, there also arose issues of divided authority and questions of responsibility, yet, the whole process was kept moving ahead. This was important for several reasons. One was that the grown-up version of Big Bertha was intended as a centerpiece for a major event in the spring of 1933 — nothing less than the General Motors exhibition at the Century of Progress Exposition in Chicago. That May, Kettering wrote that "things are apparently taking a little turn for the better around here and we are getting straightened around now with some of our work, which is going to the World's Fair at Chicago."

On May 11, 1933, a pair of eight-cylinder, inline diesel engines was shipped from the Winton plant to the World's Fair. Designated as the 8-201, they were big and they were beautiful. They had Big Bertha's eight-by-10-inch bore/stroke dimension, the welded-steel crankcase that Winton had developed and... unit fuel injectors. Borlund recorded: "The novelty in General Motors was, principally, to successfully build such a device to give accurately metered, variable quantities of fuel injected under high pressure, the Research shop being the sole producers of such devices for some time. The use of materials such as nitrided steel also promoted success by GM where others had failed."

Everyone in the corporation who had an interest in engines had some understanding of the challenges faced in developing the unit injectors. In his book, Sloan mentioned the "unit fuel injector whose parts fitted with a clearance of 30 to 60 millionths of an inch and an injector pump which built up pressures as high as 30,000 pounds per square inch as it forced fuel through holes 10 to 13 thousandths of an inch in diameter drilled in the injector tip."

During his speech to the Newcomen Society in 1945, George Codrington noted of the unit fuel injector: "By employing a separate fuel injection pump for each cylinder, the smooth, dependable performance necessary in marine and railroad service was assured. The development of this fuel injection pump is one of the major advances of the Diesel engine industry, and an entire lecture could be devoted to its development. I will only say in passing that it has clearances which are about the same as 1/120th of the thickness of a human hair.... It forces fuel through the small holes in the spray tip at a speed of 780 miles an hour."

The Fair engines were designed to run at about 720 rpm and developed 600 horsepower, although they had far exceeded these figures in tests. They weighed something over 12,000 pounds and had a power-to-weight ratio of about 22:1, a remarkable figure that was soon to be dramatically lowered. At the Fair, they were hooked up to 400 kw generators to produce electricity for GM's main exhibit, a functional assembly line producing Chevrolets.

Alfred Sloan remembered of the Fair that "we were primarily concerned with proving that Mr. Kettering's basic design was a good and practical one; we did not anticipate that the commercial applications would come as soon as they did. But before the engine for the display was even finished, our perspective on this matter was drastically changed."

The man who changed it was director of the Burlington Rock Island Railroad, Ralph Budd. When Budd learned about GM's work, he immediately visited Kettering who showed him Big Bertha but told him that the full-size, eight-cylinder version still had a long development program ahead of it. Boss was, as usual, right. Installed behind big, plate-glass windows at the Fair, the 8-201 engines required nightly maintenance to be ready for the next day's work. Kettering's son, Eugene, also an engineer, was in charge of this work and Sloan would later write that Eugene Kettering "commented that 'the only part of that engine that worked well was the dipstick.'"

But Ralph Budd perceived that, behind those glass windows lay the future of railroad locomotives. Whatever their mechanical problems, the big Winton eight-cylinder engines that he was seeing possessed the right combination of horsepower, weight, and durability to power a new kind of passenger train. And the technology for such a train already existed. It had been developed in Philadelphia by Edward Budd (no relation), a gentle man of innumerable, brilliant ideas. Budd, who had devised the first all-steel automobile bodies for the Dodge brothers in 1917, also developed a way to fuse together stainless steel panels at 2,700

GM Building, Century of Progress, 1933. Used to power the GM exhibit at the fair were these two, 600-hp, two-cycle Wintons. They gave the public its first glimpse of the modern diesel engine.

degrees. The Budd Company was building rail cars and coaches of the material, and it wanted to build a stainless steel, streamlined train that would be lightweight and low maintenance by comparison to then-typical equipment.

In October 1933, a variety of modifications were made to the third 8-201 engine to adapt it for railroad service. These changes included an exhaust system and air intake suitable for use in a new Budd locomotive, but it was understood at Winton and GM that this engine was for the Burlington railroad. It was eventually shipped to the Budd factory in Philadelphia and there installed in what would become famous in April 1934, as the Burlington's *Pioneer Zephyr.* The three-unit train weighed no more than a standard Pullman car.

On April 16, 1933, after the unit injectors had been properly adjusted, the train accelerated to 108 miles per hour over a six-mile stretch of track, fast enough to leave its tail lights behind on the roadbed. The following day, in Philadelphia, this train, the interior of which had been designed in an art deco fashion by architect Paul Cret, was officially dedicated in a ceremony broadcast nationally by NBC. Alfred Sloan was among the speakers. Now, all the many months' development work done on Big Bertha had culminated in an eight-cylinder diesel engine that was suddenly in very public view. In late May, the train made a dawn-to-dusk run from Denver to Chicago and it was reported that every grade crossing along the route was crowded with people and that, near Chicago, thousands lined the tracks to "watch the Zephyr, with its muffled roar and tiger-like smoothness, run through the twilight." The train arrived in Chicago on the opening day of the Exposition's second year and took its part in the Century of Progress as a display toured by an estimated 1.5 million people.

Thus was laid the cornerstone of GM's diesel locomotive business, which would grow to so dominate the industry that, in 1955, a Senate subcommittee was formed to examine whether GM might have used unfair tactics. Sloan wrote that, as a response to such charges, "Kettering once remarked during another congressional investigation, our biggest advantage in the locomotive industry was the fact that our competitors thought we were crazy."

Although the 201A engine was now past the research phase, many problems arose subsequently that prevented a true production model from being launched. In part, these problems stemmed from using common parts for the locomotive engines and the marine versions. But they also stemmed from ongoing issues between Winton and the GM Lab. In part, this appears to have grown from the fact that Winton personnel never quite gave up the idea of a four-cycle engine although the GM Lab and Kettering were firm believers in the two-cycle, because of power-to-weight benefits, among other things. In 1933, with the 201A having made a major impression on all who had seen it, Winton's George Codrington received a letter from an engineer in New York regarding an ad for a new yacht powered by the "latest" Winton four-cycle diesels.

"In view of the two, 2-cycle Diesel engine-generator sets that are producing the electric current for the General Motors Building at the Century of Progress," wrote engineer John Muhlfield, "will you please advise why you are continuing the four-cycle principle in these new yachts?"

Codrington responded: "... the reason is that we have been so busy with our Navy and Railroad work on 2-cycle engines direct connected to generators, that we have not had an opportunity to devote time to the designing of a direct reversible marine type of engine. We have no intention whatever of dropping our 4-cycle line. The 2-cycle is simply an addition and we will continue to manufacture both types of engines."

Kettering had a somewhat different answer when Muhlfield wrote to *him*. "We have, of course, all of the equipment for making four-cycle engines of the heavier type and when orders are received for this type of engine they will be filled. We have a wide range of piston sizes in the four-cycle, while up to the present time we have only one size, the 8 x 10, in the two-cycle."

Codrington was talking about *developing* four-cycle diesels and Kettering was talking about *taking orders* when received. Things deteriorated enough in fact, for Alfred Sloan to call Kettering to say he was planning to kill the two-cycle diesel program. Remembered Boss years later, "He called me up one day and he said, 'Ket, we've got to throw this thing out.' I said, 'Alfred, you shouldn't throw it out until you let us make one engine the way we know it ought to be made'; and I said, 'You're not fair if you don't do that. You haven't got a man in Winton that wants to make a two-cycle engine so it will run. Now, you ought to let us.' That's when I got the Winton engineering department all brought over here and we designed an engine down here, which is the engine that's running on the locomotives now." (The engine was the 567.)

Dravo Corporation Shipyard, Ohio River, circa 1944. Here, a 567 model with reverse gear is being lowered into an LST.

With the OK to keep working on the two-cycle, Kettering and the Lab refocused on other models that were far more private than the eight-cylinder installed in the Pioneer Zephyr. Even as development of the 201A engine continued for railroad applications, GM was building a series of engines ordered by the Navy. Those 6,000-cubic-inch engines, with a 60-degree, V-12 configuration, were going into submarines, and they were a special interest of the man who had done much to lead the Navy's adoption of diesel engines, Admiral S. Robinson. He had, in 1931, set forth two fundamental requirements for the future diesels to be used by the Navy. These included a power-to-weight ratio of no more than 27.5:1 and a design that could be manufactured on a production line. In March 1931, Admiral Robinson solicited bids for such an engine and by December, he had authorized contracts for five different companies to build a test diesel.

As prepared for shipment to the Navy, probably in mid-September 1933, the Winton-built V-12 was not truly ready to perform at its best. Whether this was due to issues involving preparation of the World's Fair engines at the same time, or because of ongoing doubts about George Codrington's commitment to the two-cycle is somewhat unclear. Shipment of the new diesel was postponed until engineers from the Lab had shown that changes were needed in the exhaust manifold and exhaust valve timing to improve compression and performance. It was not until the end of October that the Navy's engine was ready for shipment.

Vice Admiral Harold G. Bowen was a participant in the tests of the experimental submarine diesels and, looking back on the tests a decade later, he remembered the 12-201 in considerable detail: "Its features included exhaust valves in the head, uniflow scavenge, unit type of injector, combination fuel pump and injector, and a welded steel housing. After subjecting this experimental engine to exhaustive tests at the U.S. Naval Engineering Experiment Station at Annapolis, the engineers at the bureau, including myself, recommended it for further development and the chief decided to go along with it spite of the unfavorable report submitted by the Naval Experiment Station. For once it may be recorded by historians that a bureau was right.... The brilliant subsequent service of the model 201-A in the Navy fully vindicated its adoption."

In the end, despite some disagreement, the Winton was the only one of the five engines to survive those tests. As delivered, this V-12 produced 950 horsepower at 720 rpm. Subsequent to the tests, in November 1933, GM received an order from the Navy for 16 201-series engines and spare parts, but these were to be 16-cylinder versions. Four such engines would go into a new class of diesel electric subs launched between 1935 and 1938 that included *Shark, Porpoise, Tarpon,* and *Pike.* The first engine was to be delivered by July 4th, 1934, and there were penalties written into the contract if that engine or subsequent ones were late. These were to be $50 per day for the first 15 days of non-delivery and $100 per day thereafter.

GM 12-201 at Milford, Michigan, proving ground, 1941. This is one of GM's new diesels fitted with electric drive on display at a defense-related exhibit. The fuel lines leading the unit injectors are clearly visible. GM's Cleveland plant employed 1,500 people and engine components, according to the placard, came from 178 different subcontractors. (Copyright 1999 GM Corp. Used with Permission of GM Media Archives)

Just before Thanksgiving in 1933, Boss Ket called for a conference that included Hallett, Shoemaker, and several other key managers at the Lab. Despite the successful performance of the test 12-201, there remained a variety of issues that Kettering knew must be addressed before the engine could be put in service. Now, he asked his engineers to develop specifications for use by Navy diesel specialists for inspecting, testing, and servicing the injectors. It remained to be decided at what pressure the injectors should begin spraying fuel, and the entire matter of specifications for the engine's blower assembly and testing needed to be addressed. Remaining on the agenda, too, was the troubling issue of burned pistons and excessive cylinder wear, which were then the most serious challenges for GM's two-cycle diesel developers. It was a problem that ultimately would be addressed by design changes and the development of malleable-iron pistons by the corporation's Central Foundry Division to replace the aluminum originals.

By the end of 1934, Kettering thought that the big, two-cycle engine was well-enough developed that he began contemplating refitting the *Olive K* with four of them. "As you know," he wrote to a friend, "the present engines are six-cylinder, four-cycle 14 x 16 turning 350 rpm and the two engines and auxiliaries weigh about sixty-six tons. With the new setup I have four main engines and auxiliaries weighing about sixteen tons, so we would get about fifty tons out of the weight of the boat. . . . " It was an interesting idea, of course, but it was not implemented. On July 16th, 12 days after the first scheduled delivery, the 16-201A on test at Winton cracked a piston after 89 hours of operation. The piston was replaced with a new, oil-cooled design, and tests began again, with greater success. But development of the piston, determining ideal piston clearance and oiling methods, would continue for at least two more years.

Even as work was progressing on development of these big diesels, it was also decided to apply the two-cycle principle to a smaller line of engines that would be useful for everything from generator sets to small craft. This prompted the building of two, single-cylinder engines of 3 5/8-inch bore and 4 3/4-inch stroke. One had direct injection, the other an antechamber into which the fuel was injected.

As development proceeded, Kettering set forth goals for the smaller engine. These included that it be suitable for mass production at a favorable cost while being easily adapted to different applications. It was recognized that the smaller diesel would have markets entirely its own and that, ultimately, its development and production would be served best if a separate business unit was created to oversee it. In April 1937, R. K. Evans, the GM vice president responsible for the diesel divisions, reported to Alfred Sloan that "a small group of competent men has been assigned to carry out the production planning." It was the beginning of what would be known as the Detroit Diesel Engine Division. William Crowe, an engineer working for Kettering, was reassigned from the Lab to this new project.

"I was told by Mr. Evans that a small group of research engineers had established themselves in space at the Cadillac Motor Company and that I, with him, would go over to look over the particular designs and discuss the formulation of the new division," Crowe recalled in a 1961 interview. "It was decided after discussion with Research and Engineering Staff that the engine should be of three sizes: namely a six, a four, and a three-cylinder."

The engine Crowe was describing would become famous as the Series 71 (71 cubic inches per cylinder) and ultimately change the entire market for commercial diesel engines. But Series 71 development involved more than merely scaling down the bigger engines. Hallett reported that "it is interesting to note it has taken five or six months of testing to get good results on these engines but this work should put us in position to make a really superior line of small Diesel engines." Before that time, however, the program would verge on being scrapped because of lubrication problems that were finally solved by the Lab's research chemists. Bill Crowe recalled, too, that finding a supply source to make the prototype spiral rotor casting for the engine's blower was a challenge.

"Another item, in this respect, that Boss Ket's ingenious way of doing things, had his finger in it, was that all of the rotor castings, initially, that we got were received from Yellow Springs, being a foundry in connection with Antioch College where accurate castings were made. Later, they were started to get us going through our initial production, then we developed other sources in order to make them in larger quantities, but the method of casting them still existed as it was originally devised by Ellis McSummers."

In 1937, the first of these engines began coming out of a new factory built on Detroit's west side. Initially, production was very limited and it was not until 1938 that some 700 engines were sent to General Motors Coach and Truck for installation in buses and to Gray Marine for marine conversion. In 1939, with

Celebrating the 100,000th GM 6-71 in 1944. Charles Kettering receives the congratulations of Navy brass at a ceremony honoring production of the 100,000th engine destined for landing craft. (Copyright 1998 GM Corp. Used with Permission of GM Media Archives)

production figures still comparatively small, the engine was supplied to Allis-Chalmers for use in crawler tractors and generator sets. Then came the war, and the Series 71 entered mass production, climbing from 9000 engines in 1941 to 62,000 in 1944. And those figures don't take into account the engines built by Gray Marine.

"One of the most interesting services rendered by the diesel engine," reported *Diesel Progress* in 1948, "was the ability of the same engine to power nearly all of the various types of landing craft. This engine was the General Motors Model 71 built by the Detroit Diesel Division. It was a six-cylinder two-cycle engine rated at 225 bhp at 2100 rpm, with cylinders 4¼ in. by 5 in. It was fitted with marine reverse gear and reduction gear. One engine gave the necessary power for a 36-foot boat; and two, three and four engines were combined to provide for the larger craft; in some cases twin screws were used. Tens of thousands of these engines were built."

Speaking of the comparative volume of different diesel engines supplied by GM to the Navy during the war, Bill Crowe noted: "Obviously, the demand for our engines of smaller horsepower than those manufactured by the Cleveland Diesel Engine Division and Electro-Motive is logical in view of the fact that of the total diesel engine horsepower commissioned in the United States Navy as of July 1, 1945, Detroit Diesel percentage was 41, Cleveland Diesel 24.6; Electro-Motive, 8.1; or General Motors of all diesel engines used by the U.S. Navy totaled 73.7 percent." In 1945, the returning servicemen gave further impetus to the engines, based on their experience with them. It was their praise, in part, that paved the way for GM in the postwar commercial market, a market it was soon to dominate.

"We always spoke of Mr. Kettering as Boss Ket," Glen Shoemaker reminisced in an oral history of his career at General Motors, "but actually, there is no indication that the Boss ever bossed anybody. He was always welcome in your office, you were always thrilled and enthused by his conversation while he was there. It was quite common for him to come in unannounced and remark, 'What do you know?' This was soon followed by a brief report from the engineer, to which he listened attentively. The Boss himself would quite often express his satisfaction, make a few suggestive questions and then take off on some personal discussion of someone he had met yesterday in New York, about this or that business affair or technical subject, often entirely unrelated to your own particular subject."

It was Kettering's way both of keeping in touch and sharing his sense of curiosity and wonder regarding just about everything. He was regarded by those who knew him best as a man possessing more diversified knowledge of different scientific disciplines than anyone they had ever known. He could, in short order, captivate a listener with his stories, even if the audience had previously no interest in what Boss was talking about. This was a skill that caused Kettering's nomination to GM's policy committee to be rejected. Alfred Sloan said that "Committee members feared that 'Boss Ket's' fascinating way with words and his visions of the wonders of tomorrow would sidetrack the business of the day."

But it was that same vision that kept the Research Laboratory on track during all the difficult years of developing the unit-injected, two-cycle diesel engine. If Kettering always was skeptical of theoreticians, he revered practical experiment and development, and he always pushed the staff to keep on trying. Among his store of favorite quips was "It doesn't matter if you try and try again and fail. It does matter if you try

Gray Marine diesel as installed in a 36-foot Landing Craft Vehicle/Personnel (LCVP). (Courtesy, U.S. Navy)

and fail, and fail to try again." And that is how the GM diesels were developed. The advent of World War II served to focus what was achieved in a way that nobody connected with the diesel program would have wanted. Yet, it was remarkable.

"When all the war records can be told," wrote George Codrington in a 1943 booklet outlining the history of GM's diesel program, "the part played by the Diesel Engine will command attention, both as a great public service and as a noteworthy advancement in a highly essential industry. The record will be found significant, not only because Diesel Engines solved the propulsion problem for so many of the Navy's new fighting craft, but because their wartime service already indicates that Diesels will aid and accelerate the re-conversion of American industry in the change conditions inevitable in the postwar world."

It was, all in all, an amazing demonstration. Kettering and his lab staff, and those at Winton, had managed to overcome a bewildering array of problems to develop viable, serviceable, powerful, lightweight diesel engines suitable for volume production. And this was accomplished on a budget that was estimated to be well under the $5 million that Electro-Motive's Harold Hamilton had predicted would be needed. "Engineering," Boss Ket liked to say, "is a combination of brains and materials. The more brains, the less materials."

Kettering emerged from the intense work done during the war years with as much energy as ever but in 1945, his wife, Olive, who had financed Boss Ket's departure from National Cash Register with $1,500 that she'd earned giving piano and organ lessons, died of cancer. The loss was a terrible one for Kettering whose two lives had been his wife and the research lab, and he came face to face with the recognition that medical research still had a long way to go to defeat the disease that had killed Olive. He donated $4 million to cancer research, the beginning of the Sloan-Kettering Institute for Cancer Research.

In 1947, Charles Kettering retired from General Motors, remaining as a busy consultant and continuing his experiments with a high-compression (12.5:1) gasoline engine, and investigations into magnetism, the nature of glass, and photosynthesis, while also devoting more time to his foundation. Among the programs he sponsored was the extension of practical, cooperative education — a long-time interest of his — from engineering schools to liberal arts colleges. In November 1958, while attending a retirement dinner for a GM colleague, he suffered a stroke and died shortly thereafter on November 25th.

In 1964, as part of an oral history project, retired GM engineer Cyrus Osborn, who had known Kettering for many years, reflected on his association with Boss Ket. Osborn told the interviewer, "Well, Ket — he was a personality. When the good Lord cast him, he broke the mold. He could be as charming and breathtaking in instances as any man I've ever known, and he could be difficult, too! He never hesitated, regardless of who it was — any of his associates, Mr. Sloan, or anyone — if he disagreed, to express his disagreement in very forceful terms." But, like all who knew Kettering, what Osborne cherished most was personal. "Entirely apart from all of his contributions to the human race and to General Motors," said Osborn "his contributions to me and my family in a personal way, just through the contact, was a very, very great one. One which, as my wife has often said, is one that you can't measure."

U.S.S Porpoise *(SS-7), circa 1906. Flying a 13-star "Boat" ensign,* Porpoise *is being steered by one of her officers through a small swell, probably in Narragansett Bay near the Newport, Rhode Island, training station. In 1908, the gasoline-powered boat was shipped to the Philippines and renamed A6. (Courtesy, Naval Historical Center)*

Chapter 5

Navy

"Was your boy one of the first to land in Africa or at Salerno? Or did he leapfrog up the New Guinea coast to take the Japanese by surprise at Hollandia? Or was he one of those gallant fighters who swarmed the shores in the Marshalls or at Normandy? If so he was put ashore by those who command the Diesel powered landing craft of the Amphibious forces."

—J. H. Maloney, Advertising Manager, Detroit Diesel Engine Division, in an address to the Society of Automotive Engineers on August 24, 1944

In 1905, when the Standard Motor Construction Company in Jersey City, New Jersey, was at work on a series of steel torpedo boats for the Russian government, a few engineering-minded officers in the U.S. Navy were beginning to seriously examine the potential role of internal combustion engines. Such machines were familiar to those involved with submarine development but the idea of replacing a steam launch or steam-driven torpedo boat had yet to get beyond the talking stage.

According to critics, a 60-foot Navy launch of that era was driven by a 52-horsepower steam engine that weighed 7½ tons and took up so much space that the boat could only carry eight or 10 men. It was a situation that might have been overstated because surviving specifications do indicate greater capacity. Still the subject of internal-combustion engines was one that a forward-thinking officer could not help but examine and in November 1904, Navy Lieutenant Arthur T. Chester wrote a letter about it to the Navy

Department. He noted that launches of this type were so heavy "as to make it dangerous to put them overboard except under the most favorable circumstances, and even when afloat, so much time would be consumed in getting up steam as to make the boat inoperative for immediate service." He pointed out that a 75-horsepower gasoline engine would weigh some 1,800 pounds and require half the space of the steam engine. Within a year, the Navy had actually designed and built a gasoline engine suitable for launches, and tested it with at least some success at the Navy Yard in Norfolk, Virginia.

This did not mean that the prediction of a naval architect friend of Chester's came true. He had suggested in 1906 that steam launches would be no more than a memory in the Navy in five years. Chester's ideas were well-received by the Secretary of the Navy, Paul Norton, who said that he supported a switch to liquid fuel. But when Norton resigned his position, the matter was dropped. Thus, one J. P. Sasse, who had been involved in aspects of submarines for the Navy was, in 1911, *still* pointing out the benefits to be had by switching from steam to gasoline. He noted additional interior room and a much greater radius of operation while acknowledging in an *Army and Navy Journal* article that "there will be many who will ridicule the idea or suggestion of adopting so radical a reform. . . ." Sasse blamed the slowness in adopting gasoline power on "unreasonable prejudice against gas engines, due in many instances in the past to sad experience with gasoline engines in small dories and other craft." He suggested that having trained operators running the engines would do much to improve the situation.

There was, in fact, a reason other than "prejudice" that was recognized as being a point against the gasoline engine — the danger of explosion. There were a variety of suggestions regarding this literally volatile issue. One was that a thorough understanding of the engine and the fuel, together with proper location of the fuel tanks, would limit gasoline's dangers. It was also suggested that alcohol could be used as fuel once compression ratios had been suitably increased and carburetors and intake valves were adapted. This proved to be farfetched and, while the idea of using alcohol as fuel never developed, the concern about gasoline's potential for explosion always persisted. It was one of the reasons that the Navy began looking seriously at oil engines and diesels for small craft and large vessels, in addition to submarines.

Although progress toward adoption of gasoline engines was slow, the Navy was making some progress. Typically, this appears to have then involved testing of motors that might meet developing Navy standards for performance and durability. In 1909, for example, the Navy Department ordered a six-cylinder, 60-horsepower engine from the Clifton Motor Works in Cincinnati for testing. Clifton was already supplying engines to U.S. Engineers' offices for use in government boats on the Mississippi River system. The engine ordered by the Navy had force-feed lubricators, a Schebler carburetor, magneto ignition, an air pump, and an extra-large bilge pump. Before accepting delivery of the Clifton for testing, it was reported that the Navy sent inspectors to the manufacturer to verify that all special requirements had been met. The engine, when delivered, was subjected to what Clifton referred to as "a hard test lasting for one entire day. It was accepted as it fulfilled the entire test." This resulted in two orders, although others may have followed. Such tests were undoubtedly conducted at other Navy yards, including the one at Mare Island, which had been examining big, Union gasoline engines for several years. By 1912, the Navy was ordering apparently significant numbers of Sterling heavy-duty models, for the Buffalo, New York, manufacturer noted that such cruisers as the *Birmingham* and the *Salem,* were equipped with Sterling-powered tenders. So was the Herreshoff-designed

50-foot admiral's barge for the battleship *Louisiana.* It carried a pair of 45- to 75-hp Sterlings and could achieve 19 mph.

In addition to outright purchases from different manufacturers, the Navy either contracted for exclusive engine designs to be built at its own facilities, or acquired rights to manufacture. A few surviving

Navy Type GG Gasoline Engine, circa 1914. This two-cylinder, two-cycle engine was designed and built at the U.S. Navy Yard in Norfolk, Virginia. The engine appears to be quite sophisticated, most likely a three-port type, the mixture fed by an updraft Schebler carburetor. Current for the make-and-break ignition system is supplied by a magneto driven by an inclined shaft fitted with spur-type gears. The engine had removable cylinder heads and a water-cooled exhaust manifold. The photograph was one of several taken, in part, to explain the Navy's progress to the House Naval Affairs Committee. (Courtesy, U.S. Navy)

Testing Department, U.S. Navy Yard, Norfolk, Virginia, circa 1913. The scope of Norfolk's production is evident in this view of completed engines awaiting test runs. One-, two-, three-, and four-cylinder engines are all on view. (Courtesy, U.S. Navy)

photographs, for example, reveal a number of the engines once produced at the Norfolk Navy Yard. The earliest of these seem to have been a two-cycle, two-cylinder labeled the GG and a four-cylinder version, the KKW. The test facility at Norfolk must have been a busy place, for photos show dozens of engines lined up to be checked.

Despite the concerns about gasoline, the Navy had established a gasoline engine training school by 1915. By the time the Grand Fleet's Sixth Battle Squadron sailed for England in 1917, the battleships carried motor boats. The British were reported to have gasoline-powered patrol boats but the various admiral's barges, captain's gigs, and other small craft were still steam-driven, and big rowing gigs also predominated. According to Lewis Freeman, who reported on the arrival of the American fleet, its motor boats made a big impression. "It was the way in which the smart and speedy American motor craft slashed and reslashed Scapa Flow with their billowing wakes, cutting circles around the steamers and barges of the British battleships that awakened the greatest admiration. It was not only the flashing speed of the motorboats, but also the way in which they were almost instantly ready in emergency that aroused the envy of the crews of the steam-craft. . . . The Yankee boats were usually launched and landing men on the beach before the British boats had steam up. I recall especially Admiral Sir John de Robeck's observing what incalculable value craft of such speed and handiness would have been to his Fleet at Gallipoli."

Years later, in 1936, Freeman published an article in *Motor Boating* in which he reported that the Navy's gasoline-powered small craft ranged in length from 26 to 50 feet and in speed from eight to 18 knots. The finest and fastest of the boats, he reported, "are powered with Van Blerck J8 150-horsepower motors. Sterling 145-horsepower motors are used in the fifty-foot officer's passenger boats and the thirty-five foot gigs. [Smaller boats were powered by 20-horsepower Buffalos built under license at Norfolk and by a 20-horsepower Navy model, the KK.] Although all of these motors have given splendid service the present plan calls for the use of the V-type Navy in all new boats." It is worth noting that the passenger capacity of those 50-footers was *50.*

Navy Type KKW, 1914. The KKW appears to be a four-cylinder version of the Type GG. Mounted on the starboard side was an oil reservoir with lines leading to the main and piston pin bearings. (Courtesy, U.S. Navy)

Mark 2 26-foot Motor Whale Boat, Seattle, September 1944. Built by the Grandy Boat Company, this new whaleboat is shown with its canopy in place and two yard workers aboard. These round-bottom, carvel-planked boats were designed to transport personnel and for use as a lifeboat. Power was supplied either by a two-cylinder, 25-hp Fairbanks-Morse 45B4 1/8-2-NM or a Navy Type DA-1, four-cylinder, 25-hp diesel. Maximum capacity was 22 people. (Courtesy, Naval Historical Center)

50-foot Officer's Boat, 1939. This boat, destined for the U.S.S Wasp, was photographed at the Norfolk Navy Yard. With only the crew and a light load of fuel, the boat rides well above her waterline. (Courtesy, Naval Historical Center)

The accompanying chart, based on information prepared under the direction of Admiral William V. Pratt for Lewis Freeman's article, gives a portrait of the engines and small boats in general use in the period leading up to the mid-'30s. Even as the listing was published, however, it was about to become outdated because the Navy had finally found a way to replace the gasoline engine with a practical, small diesel.

This effort actually dated back to 1908. Among the earliest proposals on record for replacing a steam launch's engine with a diesel was put forth by Edward Meier, then chief engineer at Adolphus Busch's American Diesel Engine Company in St. Louis. Meier suggested replacing the steam engine in a 50-foot launch with a two-cycle diesel. This letter was sent to the Engineer in Chief of the Navy, Admiral John Barton, but no order resulted, perhaps because the engine that Meier had to sell itself weighed 5 ½ tons.

In 1929, the Navy purchased two small diesels and installed each in a 50-foot launch. One of the engines survived the testing but was deemed too heavy. Three years later in 1932, another diesel was purchased but again was judged too heavy to meet the 27.5:1 power-to-weight ratio that had been established as a requirement. It was a perplexing problem. Cummins, which had done so much pioneering work on lightweight, high-speed diesels, still didn't meet the requirements and the Navy was also insistent on a manufacturer with a line of engines that would serve as parallel replacements for their gasoline counterparts. Spare parts were to be common and as consistent as possible among the different ratings.

Navy Motor Boats circa 1936

Designation	Engine	Speed (knots)	Crew	Capacity
50' Motorboat	Sterling 145 hp	13	4	50
40' Motorboat	Van Blerck J8 150 hp	18	4	37
35' Motorboat	Sterling 145 hp	13	4	27
26' Motorboat	Norfolk KK 20 hp	13	4	18
26' Motor life boat	Buffalo 20 hp (Navy)	8	3	18
21' Motor dory	Norfolk 20 hp (Navy)	8	2	10

Note: Motor launches designed to carry cargo or passengers ranged in length from 24 to 50 feet and could carry from 1.5 to 13 tons of cargo when not used for ferrying passengers. The larger motor launches, capable of being carried on a battleship, were powered with the Norfolk-built V-type, a 100-horsepower engine. The smaller boats were powered by the 20-horsepower Norfolk Type KK.

Whatever the length or power of the motor boats and motor launches, a uniform draft of three feet was maintained.

40-foot Motor Launch, November 1944. This new launch — built at North Pacific Shipbuilding Company in Anacortes, Washington — is making her allotted eight knots for the camera. These big launches were powered by a Navy Type DB, six-cylinder diesel, and could carry 75 men, including the four-man crew. (Courtesy, Naval Historical Center)

Diesel engines, Norfolk Navy yard, circa 1935. Here we see workmen assembling six-cylinder Buda diesels equipped with the Ex-Cell-O fuel pump. (Courtesy, U.S. Navy)

Finally, in 1934, the Buda Company introduced a series of small diesels that promised to meet all of the Navy's requirements. The Navy acquired the rights from Buda to build the engine in three sizes, a 25- and a 60-horsepower four-cylinder, and a 105-horsepower six. The latter had the highest rated speed of the three, 1,885 rpm, and was lighter than the gasoline engine that it replaced. These diesels, built at the Norfolk Navy Yard, used fuel-injection pumps of American design manufactured by the Ex-Cell-O Corporation in Detroit, rather than the Bosch pumps used on Buda's own models.

Thus began what was known in the Navy as "dieselization," a plan to convert virtually all motorboats to diesel. This program seems to have been formalized late in 1933 after a development period that was referred to simply as "a long time." But this was going to be a *big* project. In December 1933, *Motorship* magazine reported that the "magnitude of this undertaking may be realized when the number of such boats is considered. There are approximately 800 boats with engines of 25 b.h.p., 400 with 65 b.h.p. and 400 of 115 b.h.p. To this may be added about 100 miscellaneous boats, such as torpedo retrievers, with engines up to 180 b.h.p."

Four-cylinder gasoline engine, circa 1925. By the mid-'20s, the Norfolk Navy Yard was in full production of both four- and six-cylinder gasoline engines. (Courtesy, U.S. Navy)

As gasoline engines were due for overhaul, they were now replaced by the Buda four-cylinder designated "DA" or the six-cylinder DB or DC. Sometimes, a brand new gasoline engine already on hand for installation was instead sold. In its place went a "Norfolk diesel." As dieselization took hold, "surplus" gasoline engines could represent a great find for the fisherman or commercial operator who had access to them, and some survive today in the hands of collectors. It's worth noting that, despite efforts at Norfolk to standardize its gasoline engine lineup, other gasoline engines besides the Sterlings, Van Blercks, and Norfolk-built models were used, too. On the West Coast, it is certain that the Navy purchased engines from Hicks and perhaps from Union and others, too.

As things developed, the plan to dieselize the Navy was not about to be completed in a brief time-frame. Not only were budgets tight but World War II would stretch production capacity for all sorts of engines to the limit, and gasoline engines played a role in powering a wide variety of small craft as the accompanying chart reveals. They all did yeoman service but one of them, in particular, became associated with the duty that the public perceived as more glamorous than most, for it had been selected to power the Navy's "mosquito fleet" of Patrol Torpedo boats. It was an engine whose origins could be traced back to Gold Cup racing in the early '20s, an engine that would occupy the wartime careers of thousands of the Navy's motor mechanics. It was the Packard V-12.

Errata

Page 124 line 5/6: Packard's *former* president
Pages 124-25: Edward Deeds, not Edwin
Page 145 caption: Engine is jump spark not make-and-break
Page 153 chart: The Packard 4M-2500 developed 1,350 hp @ *2,400* rpm
Page 156, line 3: 2,500 *cubic inches*

Post-Dieselization Gasoline-Powered Small Craft

Boat Type	Engine/No. Cylinders	HP @ RPM
17' Line handling boat	Gray Marine 4-22/4	26 @ 1,800
19' Rescue Boat	Lycoming 0-360-A1A/6	160 @ 2,700
24' Personnel Boat	Chrysler M7/6	90 @ 2,800
36' Picket Boat	Chrysler Royal M-8-M/8	115 @ 2,800
36' Picket Boat	Hall-Scott 168/8	180 @ 1,900
45' Picket Boat	Hall-Scott Invader/12	250 @ 2,100
45'/63' Rescue	Hall-Scott/12	450 @ 1,800
Patrol Torpedo	Packard 4-M-2500/12	1,350 @ 1,300

Whatever its potential for blowing up, the gasoline engine was recognized by naval engineers as offering so much promise for torpedo boats that experiments with such craft were under way in different countries at least by 1906. In England that year, Thornycroft demonstrated a steel, 40-foot boat that carried a five-meter-long Whitehead torpedo. With a beam of just 6.3 feet and powered by a gasoline engine, this boat could achieve 17-mph speeds. This was fast enough to suggest to U.S. authorities that, as rear admiral C. M. Chester wrote in 1906, "We have the means of doubling the power and halving the weights, by the use of explosive engines.... it is high time we reorganize our boat fleet."

To men like Admiral Chester, the advantages of the gasoline engine were too many to be ignored. In terms of a torpedo boat, a gasoline engine meant reduced visibility to an enemy because there was so much less smoke than that produced by a steam engine; the number of men handling the engine could be reduced, making more personnel available to handle weapons; speed could be controlled from the bridge; some of the weight saved by the engine could be put toward added armor or armament; the radius of action could be extended because of more efficient fuel use. It was also acknowledged that, if gasoline had explosive qualities, the danger posed by a broken steam pipe or burst boiler was no less deadly.

As aircraft engine development continued during World War I and especially in the postwar years, the potential evolution of such powerful, lightweight engines into marine versions became more practical. Liberty, Curtiss, Wright, and Hispano-Suiza engines were converted for marine use either by specialty companies or wealthy raceboat owners. But, as exciting as they may have been, such engines did not possess the stamina for prolonged, heavy-duty performance. Neither did the Packard engines developed for Gold Cup and Sweepstakes races. All these engines were operated in a tightly controlled, generally freshwater environment with mechanics aboard and tear-down facilities handy. In the end, however, two engine builders set out to methodically adapt their aircraft-related experience to a line of marine engines. These were Hall-Scott in California and Packard in Detroit, the same two companies that had, during 1917, shared experience and resources to create the Liberty engine. During World War II, both supplied engines to the Navy, Hall-Scott for rescue boats and Packard for torpedo boats. "These small trim crafts with their darting swiftness have, as this is being written, covered themselves with glory," wrote Packard's Jesse Vincent.

According to Vincent's own memoir of the PT boat's engine, Packard decided in the late fall of 1938 to pursue development of its 2500 model aircraft engine for marine use. Among other things, such an engine would give the company a product that was not subject to the seasonality of the automobile business while also, perhaps, offering the chance for some volume production thanks to the deteriorating conditions in Europe. Packard secured a contract from the Navy's Bureau of Ships to build seven experimental marine versions of the 2500, which it designated the 3M-2500.

Placed in charge of this development was Packard's George Harold Brodie. Born in Pittsfield, Massachusetts, in 1887, Brodie had joined Packard in 1904 as a draftsman and remained with the company until his retirement in 1955, after the merger with Studebaker was complete. So, Brodie was witness to Packard's greatness as well as to its decline. Among other things, Brodie had worked on the company's radial diesel aircraft engine. The engineers working with Brodie were Marsden Ware and M. J. Steele and, almost immediately after they began digging into the project, they made some unhappy discoveries. For one thing, the assumption that existing patterns could be used proved incorrect. The patterns were found to be either damaged or too difficult to modify for needed changes. In addition, the stock of 2500-model engine parts was discovered to contain defective components. Packard found itself investing an unexpected $300,000 to lay the groundwork for its new marine engine.

Packard 3M-2500, 1939. In April, 1939, Packard received a contract to build seven engines for the Bureau of Ships Experimental PT Boat Program. Designated 3M-2500, these experimental engines were quite different from the production model. On the latter, the 3M's troublesome battery ignition system was replaced by a magneto; the 3M's two Stromberg carburetors mounted above the intake manifold on the pressure side of the supercharger were replaced by one Holley mounted on the suction side; and the right- and left-side cylinders of the 3M were replaced by interchangeable cylinders on the 4M. (Courtesy, PT Boats, Inc.)

There was some good news, too. Packard contacted Hubert Scott-Paine of the British Power Boat Company, which was negotiating with the Electric Boat Company for rights to build the torpedo boat that Scott-Paine had developed. Scott-Paine was interested in the possibility of using the new Packard engine in the boat, which was then powered by a trio of Rolls-Royce-built engines. On December 4, 1939, Packard received a contract for 81 engines for Elco PT boats.

"At this point," Jesse Vincent wrote, "the Management decided that our Marine Engine program was assuming proportions which justified diverting such of my time as might be necessary to actually taking charge of the preparation of specifications and drawings to govern the construction of the engines under the new contract. . . . I was naturally governed by my own past experience with Marine engines as well as by recommendations from Messrs. Ware, Steele and Brodie, who had had the benefit of test and operation experience with 3M-2500 experimental engines."

Having experienced a variety of difficulties with the 3M experimental engines, Vincent, Brodie and the team determined that extensive redesign and retooling would be necessary to make the engine practical for production and for marine use. The revisions were so extensive that the engine, as produced, received a new designation, the 4M-2500. Some of the changes appear in the accompanying chart.

Key Modifications of Packard 3M-2500 to become the 4M-2500

- Provision for driving an air or hydraulic pump
- New, sturdier camshaft with needle-bearing rocker levers for each valve
- Single, heated Holley carburetor mounted on the suction side of the supercharger instead of two unheated Strombergs on the pressure side of the supercharger
- Fork-and-blade rather than articulated connecting rods
- Revised crankcase suitable for the Snow & Petrelli reverse gear
- Crankshaft thrust bearing revised and number one main bearing drilled to provide oil feed through to the reverse gear
- Cylinder skirts 5/16-inches longer
- Higher compression ratio piston (6.4:1 vs. 6:1) with four rings instead of three
- One-inch-smaller diameter exhaust manifold with the end cover cored for water circulation
- New intake manifold
- Larger diameter valve stems
- Supercharger impeller mounted on ball- rather than plain-bearings
- Addition of a crankshaft vibration damper
- Replacement of 3M's two water pumps by a single pump and revised water piping to simplify the engine and improve its appearance

The first 4M-2500 was completed and ready for testing on March 22, 1940. With a 6⅜-inch bore and 6½-inch stroke (approximately 2,500 cc per cylinder), the big V-12 was rated at 1,250 horsepower at 2,500 rpm. Later versions were rated at 1,500 horsepower. A series of laboratory tests was conducted on the first engine and it was then subjected to the Navy's 24-hour endurance test on May 12th after which the engine was torn down, inspected, and then approved for production. The first production engine was delivered to Elco's plant in Bayonne, New Jersey, on June 15th, 1940. In his brief memoir on the engine's development, Jesse Vincent was careful to note that Packard had rejected the use of the reverse gear, vee-drive, and salt-water pump patented by Scott-Paine.

"We decided that we were not interested in this arrangement and that we would prefer to develop an American source. Elco had difficulty in arranging for production of these units in this country and in addition it became apparent that it would be necessary to redesign them to meet the requirements of operation of our higher-powered engine. In the meantime, working with Snow & Petrelli, we had been able to obtain Navy approval on an enlarged design of a reverse gear which they had developed in its original form for the Bureau of Ships. This new design, which is our present reverse gear, was accepted by Elco who arranged for the purchase of the first 81 units from Snow & Petrelli."

The gear performed well on the dynamometer but not in a boat because of problems with clutch engagement and clashing. Packard did the engineering necessary to fix the problem but the gear was not improved to its final stage until Packard also began manufacturing it as well as the engine. By April 1945, Packard manufactured and shipped 12,115 of its 44-2500 engines. "It is a matter of pride to Packard Engineering," wrote Jesse Vincent, "that no boat has ever waited for a marine engine – convincing proof that notwithstanding the many problems involved, each new stepup in production was met. The awarding of the coveted 'E' pennant with three stars symbolic of continued production excellence and achievement further attests to the quality of the job Packard has done."

As the Packard V-12s began leaving the factory on Detroit's East Grand Boulevard, they would occupy the wartime careers of thousands of PT boat crews and support personnel. Packards powered the boats in which General Douglas MacArthur made his escape from the Philippines and John F. Kennedy's PT-109. There remain today many

Packard Engine School, Detroit, Michigan, February 1941. Big and beautiful, this 4M-2500 is the object of study for a group of technicians. Both civilian and Navy personnel were trained at the school. (Courtesy, PT Boats, Inc.)

veterans of the PT fleet who, as young men, were assigned the task of keeping the big V-12s operating, quite often under the difficult circumstances always attendant to a combat environment. Among them was Don Mason who, after graduating from technical school, got a job at the Packard plant in its marine engine division. This proved short-lived thanks to issues relating to the union and seniority, but Mason and five others were given the chance to attend the 16-week marine engine school that Packard established to train mechanics on the 4M-2500. In April 1942, as a civilian technician with the Navy, Mason found himself in Alexandria, Egypt, at the HMS Mosquito base, home to 20 Vosper and 20 Elco torpedo boats.

"There were no spare parts available," he said, "so mostly we did troubleshooting. There were problems with the supercharger impeller and the vibration damper, which was OK at most speeds except for 1,550 rpm. We begged, borrowed and cannibalized but we had a good record and never had a failure that cost anyone their life. The boats did coastal patrols, raided Tobruk and carried commandos."

Before he left the Mediterranean, Mason was stationed in Haifa, Beirut, and Syria, working with men from Canada, New Zealand, and Britain, as well as the United States. Then he was assigned to the Pacific where he worked at bases in the New Hebrides and the Philippines. "In the Pacific," he said, "we had plenty

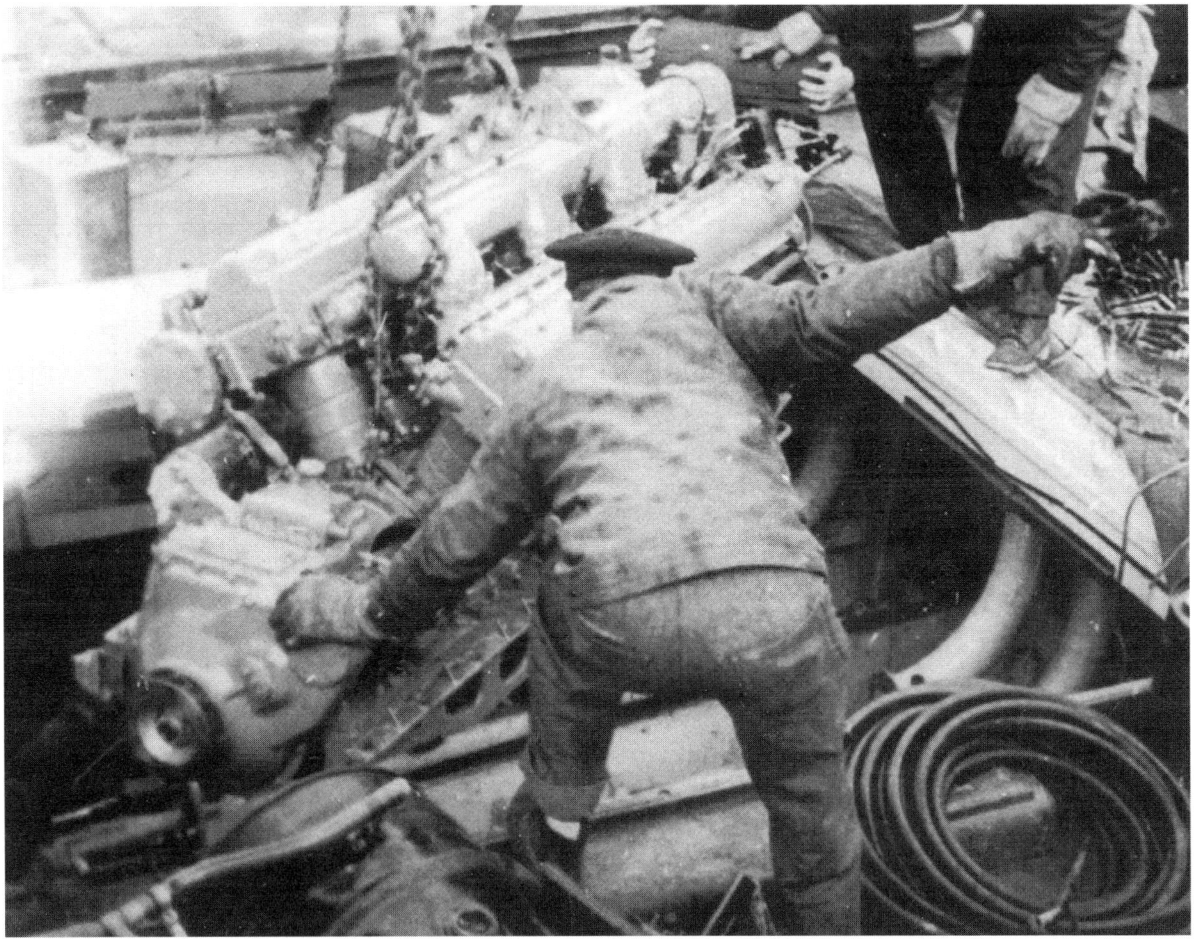

Portsmouth, England, circa 1944. Skilled mechanics kept PT boats operational. Here, a Packard is being lowered into the engine room. (Courtesy, PT Boats, Inc.)

of parts. The engine ended up being very dependable and could go a thousand hours. The 'motor macs' – there were three per boat — liked the Packards real well. Really, it was the best engine they could find for that use. Conversions of the Rolls-Royce Merlin were tried but wouldn't stand up to the loads. They weren't rigid enough."

It was a war story that, all in all, probably exceeded Jesse Vincent's expectations because the Packard engine became something of a legend in its own time. Vincent, writing in 1945, likened it "to an 'unsung hero' of World War II." But over a half-century later, long after the unfortunate demise of Packard itself, the legacy of the 4M-2500 marine engine remains very much alive.

Considering the doubts raised within the Navy about the safety of gasoline engines for use in a launch, it may today appear shocking that the earliest *submarines* had gasoline engines. In his initial submarine designs, John Phillip Holland connected his fascination with submarine efforts like the Confederacy's *Hunley* and news from Europe with his interest in the internal combustion engine developed by George Bailey Brayton in 1872. He built four gasoline-powered boats in the 10-year period between 1878 and 1888 and then won a Navy-sponsored competition to design a little submarine for coastal defense purposes. This project, however, became enmeshed in politics and bureaucracy. Instead, he began developing a new boat, *Holland VI*, which garnered the support of Theodore Roosevelt, then Assistant Secretary of the Navy and later, Admiral George Dewey. This boat was commissioned by the Navy as the *USS Holland* on October 12, 1900, and the Navy contracted for seven more Holland submarines to be built by Holland's Electric Boat Company.

The *Holland* was a 54-footer powered by a 50-horsepower Otto four-cycle engine. In August 1901 *The Gas Engine* magazine reported that a government inspector at Holland had approved a larger version of this

U.S.S Holland *(SS-1), 1899. Here is the Navy's first submarine, powered by a 50-hp Otto engine, underway in Long Island Sound off Greenport, New York. (Courtesy, Naval Historical Center)*

U.S.S Shark *(SS-8) and U.S.S* Porpoise *(SS-7), Newport Training Station, circa 1905. (Courtesy, Naval Historical Center)*

engine for the *Holland*'s successors. These big Ottos were of interesting specification. They were four-cylinder machines with the cylinders cast in pairs and bolted to a common base. Length was just under 10 feet and height a bit over six feet. The flywheel was 5½ feet in diameter. An exposed camshaft was mounted on the engine's starboard side, about two thirds of the way up the cylinders. It operated the side-mounted air intake valves and the overhead exhaust valves. Fuel entered through valves that could be controlled by the engineer. The fuel-air mixture was exploded by "hammer-break" igniters, and a governor-controlled engine speed by a small roller that was pushed out of the way of the cam lobe so that the gasoline inlet did not open. It was similar to a classic "hit-and-miss" stationary engine's system. The engine developed 160 horsepower at 320 rpm and 190 horsepower at 360 rpm, giving the 13,000-pound machine a then quite favorable power-to-weight ratio of 70:1.

These Holland submarines resulted in additional orders for the company by 1910, and Britain arranged a licensing agreement for five boats, too. But clearly, such boats were dangerous. "During the first ten years of submarine operations," wrote Clay Blair, Jr., in *Silent Victory,* "there were many disastrous and near-disastrous explosions on board gasoline-powered submarines. Sometimes during submerged operations, while running on the battery, gasoline fumes seeped into the hull and intoxicated the crew to a dangerous degree."

This did not stop inventor John M. Cage, in Los Angeles, from his attempts to demonstrate the ultimate practicality of the gasoline-powered submarine. In 1913, Cage built a 75-foot boat powered by two 90-100 horsepower Buffalo gasoline engines that were the vessel's *sole* source of propulsion, whether surfaced or submerged. This was accomplished by forcing the engines' exhaust out of the hull under pressure, a feat that could be accomplished at depths of as much as 150 feet. In 1913, Cage's boat remained submerged for 36 hours off Long Beach to claim what was reported in *MotorBoat* magazine to be the world's record while "the crew of six men suffered practically no ill effects from lack of air, and the mechanical equipment worked perfectly."

By that time, the Navy had at least decided that diesels were the only practical engine for driving a submarine on the surface. As described in Chapter One, Electric Boat's New London Ship and

Engine Company (Nelseco) began building, under Vickers and M.A.N. licenses, diesel-powered submarines beginning in 1911. Subsequent to World War I, the United States, in 1922, became party to an arms limitation treaty that limited submarines to a maximum of 2,000 tons displacement. The existing M.A.N. diesels were not only comparatively complex and unreliable but now were too big for the new boats. Thus began the ongoing effort to develop lighter-weight, more compact diesel engines that led in 1932 to contracts with Winton (GM), Fairbanks-Morse, and to the Hooven-Owens-Rentschler Company (H.O.R.), in Hamilton, Ohio, another M.A.N. licensee that enjoyed some success in the early '20s building big, double-acting, two-cycle diesels for several ships. Its submarine diesels, however, proved to be so unreliable that breakdowns on war patrols were typical and the H.O.R. engines were replaced by Wintons or Fairbanks-Morse units.

Duty in early submarines was dangerous and uncomfortable and it took an adventurous sailor or forward-looking officer to apply for service in the boats. One of these officers was a lieutenant named Chester Williams Nimitz, and he quickly became among the Navy's leading diesel engine experts. His abilities were recognized both in and out of the service and in 1915, Nimitz received a job offer from Busch-Sulzer. The St. Louis company offered Nimitz, who was then making some $288 a month a contract worth $25,000 a year for five years. According to E. B. Potter's book, *Nimitz,* the future admiral replied that he was going to remain in the Navy, prompting the response, "At any rate, money is no obstacle. Write your own ticket."

Nimitz was very much involved in engine design, assembly, and installation. Once, he was on a scaffold erected around a big diesel when it collapsed. Although he escaped apparent injury, he would later attribute back problems to this accident. Also, while demonstrating a running diesel to a group of visiting engineers, he lost much of his ring finger to an exposed set of gears. In 1912, Nimitz took a lead role when the Navy was authorized to build a new group of tankers with M.A.N.-licensed diesel engines. These were enormous, six-cylinder machines with 20.20-inch bores and 39.37-inch strokes that could develop 2,500 horsepower at 1,300 rpm. In 1914, Nimitz went to Germany to inspect the M.A.N. works firsthand and learn how the engines were designed and built. He converted the engine's dimensions from millimeters to inches and returned to supervise construction. In 1916, the tanker *Maumee* was launched and Nimitz was assigned to be its executive officer and chief engineer. The vessel was moderately successful but was retired after World War I because the engine was so heavy, or the hull was so light, that crankshaft alignment was an ongoing problem. The *Maumee* was later revised with a steam engine, but the practicality of a big ship's diesel engine had been proven to the Navy. By the time the United States entered World War II, General Motors had available its 278 and 278A engines. The latter were, with a 9¾-inch bore and 10½-inch stroke, available in six-, eight-, 12-, and 16-cylinder models producing up to 1,600 horsepower at 720 rpm.

Although the Navy did much to spur general diesel engine development in the interwar era, it never did lose sight of the goal of achieving ever-improved power-to-weight ratios. In 1937, Navy officers visited the GM Research Laboratories to discuss the creation of a really lightweight, high performance diesel suitable for a new, wooden, 110-foot submarine chaser that was being designed by Navy Captain A. Loring Swasey. During World War I, Swasey had designed a 110-foot sub-chaser powered by a 600-horsepower gasoline engine. By late 1937, Charles Kettering had sketched out a most unusual diesel engine that promised to offer *twice* the power of the gasoline engine in the World War I-era boats. It was a radial engine, with cylinder banks stacked one atop the other and a vertical crankshaft that drove the propeller shaft through a 2:1 reduction, spiral gearbox. The basic architecture was intended to ease access to the cylinders for servicing, which it did in fact do. As produced, the engine was composed of four banks of radial cylinders, four cylinders to a bank. The upper end of the crankshaft drove a centrifugal blower. This 16-cylinder engine was known officially as the 16-184 but it was more generally referred to, because of its stacked cylinder arrangement, as "the pancake diesel."

Now, the GM two-cycle philosophy was taken to its most exotic level. The engine developed 1,200 horsepower at 1,800 rpm and weighed about 20 percent of one of GM's 567 model locomotive diesels. No reverse gear was needed because a variable-pitch propeller was part of the program. The engine was both compact and lightweight. It weighed about 4,800 pounds and so had a power-to-weight ratio of about 4:1. The first pilot vessel was run through a preliminary acceptance trial on July 28, 1941, and a pilot production model was completed at GM's new factory in La Grange, Ohio, that October. Development work continued even as the Navy's Inspector of Machinery approved the engine for production in February 1942.

All in all, the 184 was just the sort of thing that Kettering might have devised, unfettered by conventional theory and made practical by dint of concerted effort. Those who knew Boss Ket well were uniform in recalling that among the human traits he could not abide was the tendency to assert something couldn't be done because it broke rules or wouldn't work because nothing like it had ever been tried before. In the early days of War War II, a stunning example of Boss's reaction to such thinking occurred when a British admiral sent by President Roosevelt arrived at the Research Laboratories. The Royal Navy was badly in need of a high horsepower, lightweight engine for its own subchasers but the British Admiral, upon inspecting the 184, insisted it would need to be mounted on its side because it could never be properly lubricated in its vertical configuration.

Cyrus Osborn, who would finish the war years as general manager of GM's Electro-Motive Division, was present for this discussion and heard Kettering respond that the engine had proven satisfactory and was going to be built as it stood. It was, essentially, a take-it-or-leave-it proposition insofar as Boss was concerned. "The argument went on furiously for some time," remembered Osborn, "... and finally the British admiral said in desperation, 'Mr. Kettering, aren't you interested in saving the British Isles?' Ket said, 'No, not if it means saving a bunch of stupid people like you.' That broke up the meeting in short order."

Soon, the Pancake engine was in service in operating theaters as diverse as South American convoy routes and the Aleutian campaign. Surviving field reports provide insight into how the engine was to live with: "Never had any trouble starting their engines, other ships using the same fuel had considerable trouble in starting."

"To date the only repair on engines 360 and 367 has been a leaky liner in engine No. 360."

"The port engine was used alone to escort a 17-knot transport.... The starboard engine at various times was used at speed and loads above normal.

"The two engines, No. 169 and No. 166, removed from SC1014 after 3,000 hours of service were in better condition than any engines ever disassembled regardless of their length of service. We found no parts in a condition that would not warrant at least 2,000 hours more service. The greatest gap on any compression

GM 16-184A. Popularly known as the "pancake" diesel, this unusual engine produced 1,200 hp and possessed an extraordinary 4:1 power/weight ratio. Vertically mounted, it drove a variable pitch, reversing propeller and powered a class of subchasers.

ring was 0.0059". The valves were all in perfect condition. No springs were broken in the blower coupling. No liners or pistons were scored."

Between May 1942 and May 1944 when the Navy contract for the Pancake engine had been filled, the engines were shipped from GM's factory at the rate of 20 engines per month. The variable-pitch propellers, shafting, lube oil tanks, exhaust manifolds and bridge controls were shipped ahead of time to the 32 shipyards building the subchasers, which carried the Navy designation SC. Awaiting both the engines and these various components at each shipyard was an Electro-Motive field engineer whose responsibility it was to oversee what was obviously a rather nontraditional machinery installation. These men worked in yards from Rockland, Maine, to Los Angeles, from Duluth to Corpus Christi.

Even as the Pancake engine was being developed, the Navy's Bureau of Engineering was holding meetings with GM about the need to step up production on the 567 model, essentially an improved version of the 201. GM originally intended the 567 for use in locomotives but, in 1938, the Navy ordered a dozen 12-cylinder versions and their electric motors for three big tugs named *Seminole, Navajo,* (both later sunk in action) and *Cherokee.* These orders were followed by others from both the Navy and the Coast Guard, the latter supplied with 8-567s that featured inboard-mounted blowers that reduced the engine's width. This space-saving change was later included in the 567 engines supplied for use in the Landing Ship Tank (LST).

The scope of production involving the 567 can be grasped when one considers the shipyards that produced the vessels in which it was installed. These included Dravo in Pittsburgh; Chicago Bridge & Iron in Seneca, Illinois; Missouri Valley Bridge & Iron in Evansville, Indiana; American Bridge Company in Ambridge, Pennsylvania; and Jefferson Boat & Machine Company in Jeffersonville, Indiana. But even the capacity of all these yards was not enough to meet production schedules and the LST program was extended to Navy Yards at Boston, New York, Philadelphia, Norfolk, Charleston, and to Bethlehem Steel Company yards at Quincy, Massachusetts, and Baltimore, Maryland, Kaiser Company yards in Richmond, California, and Vancouver, Washington, and the Newport News Shipbuilding and Dry Dock Company. Operating the LSTs would be a force that eventually totaled some 50,000 men.

General Motors' expansion of production to meet Navy requirements was echoed to varying degrees by other engine companies. When the Navy told Fairbanks-Morse it would need every one of its opposed-piston, two-cycle engines, the company built a new factory in Beloit devoted exclusively to its manufacture. This two-cycle was an engine whose development was begun in 1934 and represented Fairbanks' approach to building a high horsepower, lightweight diesel. It found application in both submarines and fast surface craft.

When he looked back on D-Day, General Dwight David Eisenhower referred to the slab-sided landing craft that carried the troops ashore as "the little boats that won the war." The boats had been designed and built for the Navy, after considerable interference, by a flamboyant genius in New Orleans named Andrew Jackson Higgins. In his book, *D-Day, June 6, 1944,* historian Stephen E. Ambrose recalled something Eisenhower once told him about Higgins. "If Higgins had not designed and built those LCVPs, we never could have landed over an open beach. The whole strategy of the war would have been different."

Outspoken, possessing great vision, and an independent thinker, Higgins succeeded in developing his landing craft despite a Navy bureaucracy that didn't want to be bothered by the ideas and boats of a man who appeared to be an upstart far removed from the boatyards and designers with whom they normally dealt. "In Washington he was an outsider," wrote Jerry Strahan in his *Andrew Jackson Higgins and the Boats that Won World War II,* "and perceived by many, including officers in the Small Boat Division of the Bureau of Ships, as a threat. If they could not outdesign him, they had to find a way to control him."

The Navy was then attempting to develop a Landing Boat Personnel (LCP) based on the big sea-skiffs developed on the New Jersey coast and Long Island, and later, a steel boat designed expressly for the purpose. When Higgins saw the blueprints for this proposed boat, according to Strahan, "he wrote something to the effect of 'This is lousy' or 'This boat stinks — AJH."

Higgins was not going to make many friends this way, but he did not particularly care. He had been developing a shallow-draft workboat with its propeller housed in a tunnel stern that permitted operation in the very shallow waters around New Orleans. In fact, the boat's draft was a mere 10 inches and Higgins liked to demonstrate the *Eureka*'s ability to beach itself on a Lake Pontchartrain seawall and then back off. The *Eureka,* with its spoon-bow and wooden construction, must have struck many Navy men as being as unconventional and undesirable as its outspoken creator. But there was no denying the fact that Higgins was ready and able to produce these shallow-draft boats that were seaworthy, fast enough to outrun a following sea when fully loaded, and capable of being driven onto and then off a beach.

Richard "Mac" McDerby, who had learned boat handling on the Mississippi River, was in charge of training, in New Orleans, the servicemen who would operate the landing craft. "It was Amphibious Command training all the branches," he said. "The men had to learn how to get these boats on and off a beach. Most who came to us didn't even know what a propeller was. But it turned out that the guys who knew boats always wanted to throttle back. But you had to get past the outer bar and onto the beach. The Higgins was a great surf boat. Once on the beach, you kept the engine running to dig a hole. Then you reversed and water washed out of the tunnel. You could get them back off with the boat out of the water."

With the design approved, Higgins began mass-producing the landing craft in his own factories. At first, they were powered by Hall-Scott gasoline engines. "I was training marines in Hall-Scott-powered boats in 1941," said McDerby. "But later, the Navy wanted diesel. It had better fuel consumption and less fire hazard."

It was this insistence that led to the trials of the best diesel for the job and a Higgins boat was sent to Detroit for a trial installation of Gray Marine engines, the marinized version of the GM 6-71. Final selection of the Gray Marine diesel for exclusive use in landing craft was not a foregone conclusion. Later, in New Orleans, tests were held between Gray Marine-powered boats and those equipped with aluminum block engines built by the Superior Engine Company. It was no contest, however. Gray's mechanics actually helped their competitors, but Superior was dropped from the landing boat program.

For a time, there was discussion about exactly which engines to use in what small boats but, ultimately, the Navy decided to use the Gray Marine diesel in landing boats and use the Packard V-12 in PT boats. The challenge then was to develop a standardized version acceptable to the Navy's different design sections. This was made simpler by the fact that the 6-71 was compact enough to easily permit multi-engine installations as necessary. Two, three, or four engines could be used to drive one or more propeller shafts. In the LCU(L),

Sea trials of a just-launched, Higgins LCVP. Photographed shortly before Christmas in 1943, this 36-foot landing craft is being put through its paces. The boats were powered by one six-cylinder Gray Marine (GM 6-71) diesel. (Courtesy, U.S. Navy)

this flexibility reached its ultimate, for it was powered by two "quads," units of four Gray Marine diesels, each representing 900 horsepower. All this led to what John Mulford, then Gray's president, always thought of as Gray's greatest glory. And it was how, when the Higgins boats surged toward the Normandy beaches on June 6th, 1944, a six-cylinder, 225-horsepower Gray Marine diesel was powering each one.

By the end of the World War II, the diesel engine was as much a part of the Navy as the steam engine had been before World War I. When he reflected about the diesel engine's role in the War, Admiral E. L. Cochrane, who had been the Navy's Chief of the Bureau of Ships summed things up this way: "The United

Cooper-Bessemer Type GSB, Seattle, circa 1943. This big diesel has arrived at the Winslow Marine Railway where it is about to be installed in the minesweeper AM-294. (Courtesy, Knox County Historical Society)

States, in this last war, built thousands of ships and 90 percent of them were diesel-powered. There was no vessel that was not equipped with diesel power of some kind: much use was made of diesel auxiliaries. The whole landing craft program was diesel powered, and these ships worked under the most strenuous conditions. In many an emergency, when a ship was bombed, diesel engines saved the day. Not a single one failed us in a critical operation."

Appendix A

Preservation and Interpretation of Antique Marine Engines

Nobody knows for certain where all the old marine engines that once existed have gone, but this is an ongoing topic of some speculation for those who collect and restore the machines. Most collectors have heard tales of once-magnificent engines resting at the bottom of a Midwest lake, of others hoarded by back-country eccentrics who jealously guard them for no particularly good reason. One also learns of engines that could have been bought for peanuts except that they were lost to government scrap drives. Often, in fact, when an engine does turn up, it has been stripped of its brass parts, pieces which contribute immeasurably to the old machines' aura and authenticity but which are always difficult and often impossible to replace. Instead, they've been taken for their scrap value.

This is unfortunate. These engines, particularly those from the first few decades of development, now represent nothing less than mechanical and metallurgical works of art. To behold one in good, original condition is to be in the presence of something marvelous.

It is fairly easily to surmise the typical fate of old marine engines. When replaced by newer models, most were scrapped, became mooring anchors or were left to molder in the back corners of boatyard sheds. Sometimes, old marine engines were simply set down on a riverbank and occasionally they're still there. People know about them, too. There's a Buffalo on a Mississippi riverbank, an old Atlas diesel at a once busy but now abandoned Alaskan cannery. It would be removable only by helicopter. There's a Lockwood-Ash rusting away in a steamy bayou.

What is clear today is that only a tiny fraction of the marine engines built between 1890 and the onset of World War II have survived. For those who understand and appreciate the old iron, it's all a sad story. Just ask the man who's sought a specific engine, whether it be an Imperial from San Francisco, a Sterling from Buffalo, New York, or a Van Blerck, from Monroe, Michigan or Detroit. One hears of searches reminiscent of those for lost gold mines. "A Van Blerck?" mused one enthusiast, "*Any* Van Blerck would be a treasure."

Some marine engines have been "officially" saved, of course, and they turn up in some surprisingly diverse places. There is, for example, an exceedingly rare old Daimler at the Connecticut State Library in Hartford, and an Iowa engine rather incongruously on display in a little room at a Louisiana parish library. Maritime museums comprise the most obvious resting place. There's a nice collection at the Columbia River Maritime Museum, a number of engines at the Smithsonian, at the Maine Maritime Museum, the Antique Boat Museum, and many other such institutions. Considering the fate of so many engines, each one that has survived to become an "antique" is a treasure of sorts.

Of course, it is easy to forget that, in their time, the engines were simply tools, and sometimes frustrating ones at that. The owner of a pleasure launch in 1910 would have gladly replaced a balky engine with an improved version, easier to operate. Probably, he'd have been happy to see the last of the troublesome engine, much as we would today. And the wealthy owner of a racing boat would not have been too attached to its engine to replace it immediately if something that promised more performance became available. So, although there were people who simply could not bear to part with an engine for one reason or another, the machines were seldom thought of as technological treasures to preserve for future generations.

Despite the appeal and rarity of old marine engines, the very real question of what to do with them confronts both enthusiast and museum. The engines are, for the most part, big and heavy. Unlike an antique automobile, they can't move under their own power. Men who would collect these artifacts become adept forklift operators. They usually have one or more steel buildings or good sheds in which to store their often fascinating, if rusty, discoveries. Nor is restoration simple. Usually, a collector must establish a technical support system that may be composed of specialist craftsmen throughout the country.

If finding, storing, and restoring antique marine engines is a challenge, there is also the occasionally perplexing surprise that can attach itself to the endeavor. Consider the case of a collector who thought he was purchasing at auction an original Gray Marine diesel, still sealed in its original storage drum. What emerged, however, was not a Gray but, incredibly, a supercharged, World War II-vintage Hall-Scott V-12. But what was one to do with such a big and powerful machine? Shipping it cross-country to someone interested would have exceeded the engine's value. For the moment, at least, this old marine engine will continue to sit idle, still accompanied by its original storage documents.

Why bother to collect and preserve old marine engines? The quick answer is why not? The engines are no different than the antique automobiles, the antique boats, the antique furniture, the antique folk art or the "antique anything" that winds up in museums. All these other objects, however, have a head start on old engines. The reason is simply that they are more widely appreciated by either the general public, preservationists or well-schooled dealers or historians. Each has come to represent a market and possesses a constituency that may be supported by an academic infrastructure and even a well-defined monetary value.

The field of preserving antique engines, whether marine or stationary, has yet to reach the degree of sophistication of these other objects. But it is a field that attracts an eclectic group of dedicated people both in the United States and abroad. It is well that these people are there. Engine preservation saves the technology that each machine represents and educates us in the innovations and ways of the past. The old machines foster links to the people who used them and, in the case of old marine engines, associations with a fascinating variety of boats and far-off waters. Most of all, perhaps, the engines can become a unique window from which to view a bygone way of life.

Today, while antique marine engines can be found on display in a variety of museums, much remains to be done with these existing machines and with others still to be discovered. It is already recognized in some institutional quarters that the state of engine preservation needs to be improved. Like any museum acquisition, a decision on a marine engine will be affected by considerations such as budget, display space, and staff skills and availability. Because they can seem so specialized (even though they're not), acquiring a marine engine also raises the question of the museum's will, focus, and basic curatorial knowledge.

Once an engine has been acquired, the challenges include stabilizing and maintaining the unrestored machine, which need not be a major endeavor. Correct restoration, however, represents a more significant effort that may involve a search for replacement parts, determining the correct color, and locating support literature. Most important of all, and key to a better understanding and appreciation of the engines, is sensitive and meaningful interpretation. Such interpretation needs to include three basic themes: 1) what era of technology the machine represents 2) background on the manufacturer and 3) what role the engine played in the lives of those who would have used it.

Information about the first topic increases our understanding of the state of technical knowledge at the time the engine was built. It also serves to place the engine in the context of its time and in the overall development of technology, much as one can do in a museum dedicated to automobiles or airplanes.

An understanding of the company behind the engine spurs interest in the business history of a given geographical area, and interest in the biographies of those who created the company. Finally, it may be the association of the "iron" with people going about their everyday pursuits that matters most of all. This is what humanizes the engine to the greatest degree for it presents the engine as an important tool needed to gain one's living on the water or as an example of how earlier generations adapted the engine to recreational pursuits. Even the ads for the old companies present a meaningful insight into bygone eras. It's a rare museum engine, indeed, that is ever actually run, an operation that perhaps does more to bring the old machines and technologies to life than anything else.

One must recognize, of course, that museums can have competing demands. In a large institution, direction may alter as curators shift emphasis and reallocate budgets. Elevating the sophistication of antique marine engine collections and knowledge to its rightful place may yet require years. By contrast, the individual collector remains dedicated to only one goal — finding researching, and restoring marine engines and celebrating the past that they represent. Thus, it remains individuals who still form the real backbone of marine-engine preservation. For their efforts, collectors are typically rewarded mainly by the intrinsic delights of their avocation and by their deeply held feelings about the importance of the engines and what they represent.

There is also the support, interest — admiration even — of others in the hobby. Information continues to be freely exchanged although the esoteric nature of the field leads a few individuals to become possessive, in the apparent belief that sharing will diminish some vaguely imagined monetary worth. This is more fantasy than reality. While there may be a reasonably defined market for some old marine engines, the market in general is geographically widespread but still comparatively small in terms of actual numbers. The worth of an old marine engine, even an intact and rare engine, often boils down simply to what the one man interested thinks it's worth.

Despite the challenges, there are today, in different parts of the country, museums that are doing a credible or better job with marine engines. There is also, on the part of several, the desire to do more. This is a laudable goal but, as this is written, there is not a single public museum, public or private, dedicated to marine propulsion. The focus on preserving sail or even steam-driven vessels today only serves to define the great void that must now be filled regarding internal combustion. Prewar gasoline and diesel-powered tugs, towboats, and fishing craft are in greater need of acquisition than runabouts which are in comparatively abundant supply. But those boats, too, and the racing and pleasure craft of the early years of the century, each with its unique engine, need to be saved and correctly restored, interpreted, displayed and, when practical, operated.

Perhaps the best opportunity to accomplish all this would be creation within existing institutions of a Marine Engine Hall. Upon entering, the visitor should be transported back in time to hear sounds and see sights that were common to an earlier generation but are now threatened with being lost altogether. Interactive displays could offer the opportunity to prepare an engine for starting, and even to run it. Visitors could be involved in computer-based learning experiences about the engine's technology, the history of the company that built it, the town or city where the company existed, and the sort of boat in which it was used. If the engine was used in a fishing boat, tug boat, or tow boat, the opportunity exists to have visitors gain insight into how work was done in a bygone time. All this would be well worth the effort. It remains to be seen what visionary management will take up the challenge.

Appendix B

How to Operate the Kahlenberg

The following was written by an anonymous but thoroughly expert operator of a Kahlenberg engine.

One interesting thing about the Kahlenberg engine is the fact that is has individuality. Each engine is built to a standard and it is not shipped until it has demonstrated that is is right in every detail. Engines have frequently been kept on test for several days, tuning up, after having passed the prescribed inspection, but because Kahlenberg felt that the set-up might be altered a little here and there to make it function a little more perfectly. A "production" engine would have been shipped. This personal final inspection and regard for the individuality of the engine is refreshing in these days of quantity production, where parts are rushed through operations and inspections, and assembled and shot into the market with all possible speed. Would you prefer an engine in which the individuality was developed or surpressed?

The design of the Kahlenberg oil engine places it in the "surface ignition" or "semi-diesel" class. The former term is probably the more accurate, as, in this type, combustion is produced by the injection of fuel into the combustion chamber and against a heated surface. Initial heating of this surface is obtained by torches – thereafter the heat of combustion maintains the surface at the proper temperature.

The engine operates on the two stroke cycle, compressing the air in the base, under the piston, or the down or firing stroke, by-passing this air at the bottom of the stroke to the cylinder, where, through the proper design of ports, piston head and combustion chamber, the products of combustion are entirely blown from the cylinder. On the up-stroke, this air is compressed to the required degree, and, at the proper time, under the control of the governor, fuel is forced into the combustion chamber by a simple plugger pump, one for each cylinder, without the use of compressed air. In this type combustion is complete only if the design is right. The old saying that "the proof of the pudding is in the eating" holds good in regard to internal combustion engineering just as it does in the culinary art, and Kahlenberg design is proved right by the fact that the exhaust is smokeless.

This Kahlenberg engine has only one pair of gears, one of tool steel and one of a special bronze, spiral cut, running in an oil bath. These gears drive the governor, and, although the only work done by the governor consists in shifting a small roller, the gears are six inches in diameter, two inch face, and weigh eleven pounds each. There are no valves to grind, no cams, nothing to adjust.

Come into the engine room and see for yourself just what it is to operate a four cylinder one hundred horse power Kahlenberg engine. In the first place, you will see that the main shaft bearings are not enclosed in the crank case, but are readily accessible between the wide set of cylinders. Engineers handling some oil engines with enclosed cases tell me that they never can tell when a crank bearing journal starts to warm up, and the first knowledge they have of a heated bearing is when the babbitt has started to run, and the bearing is "gone". These engineers tell me that, when such a bearing has burned out, it requires at least five days work by two men to effect repairs, and that, when they burn out a bearing during the height of the season, the loss to the owners, due to the tie-up of the vessel, frequently equals fifty percent to the full cost of the engine.

Such a loss cannot come to the owner of a Kahlenberg. There is no knowledge at the factory of a single burned out crank journal bearing on this type of engine. However, should any shaft bearing burnt out, which, of course is possible, it can be replaced with a spare bearing shell in less than thirty minutes by one man.

Then, as an extra safeguard against trouble, the main shaft bearings are water jacketed – cooled by seawater direct from the circulating pump. So, as long as lubricating oil is supplied, a burned out shaft bearing is practically impossible. The crankshaft is the "backbone" of the engine, and upon it the life of the engine is largely dependent. Kahlenberg has long been a "crank" on crankshafts and twenty-five years of marine engine building has given them valued experience.

Look at this four cylinder hundred horse engine – ample diameter of crank shaft, the end bearing 10" long and the intermediate 8¼". And an 8¼" bearing aft of the clutch. To give you another idea of the tremendous bearing area of the crankshaft of this engine, it is 665.92 square inches, or over four and one-half square feet.

Now, let us look at the bearings themselves. The shells are bronze, with the highest quality of white metal linings, sweated in, peened, and bored to an exact fit. The No. 1 bearing, for instance, weighs seventeen pounds. Contrast this, as a bearing job, with the die cast bearings used on some engines, or with iron shells lined with poured babbitt. And bearing replacement, if ever necessary, is so easy. I have seen a main bearing shell taken out, replaced and adjusted in twenty-five minutes.

Two years ago, I met a boat owner in Milwaukee. He had a three-cylinder Kahlenberg engine, sixteen years old, and he told me that no bearing replacements had ever been made. I also saw a 3-cylinder 75 HP engine in a freight and passenger boat that had not a single shaft bearing replacement in six years, and this engine had averaged over ten hours a day during that time, or, in all, over twenty-one thousand hours. I cite these instances only to show you what you can expect from bearings on Kahlenberg engines if they are properly lubricated.

After this digression, let us return to our engine. As before stated, starting is accomplished by heating a tube in each combustion chamber with torches. These torches are built into the engine, do not require pre-heating, and use the same fuel oil distillate or kerosene that is used in the engine. The easiest way to handle them is to use a small wick torch similar to that used by the locomotive engineer for looking over his engine. Light the wick of this torch, turn on the air and then the oil on No. 1 torch, and you have, instantly, a clear blue flame, playing into the torch flue and concentrated on the starting tube, which is about three-quarter of an inch in diameter and three inches long. So intense is the flame that three minutes are sufficient for starting under the most adverse conditions. You walk along, lighting up the other torches, then open the sight feed lubricators, one on the air distributor and one for the pump eccentrics. Then, using the hand plungers, shoot five or six strokes of fuel into each of the cylinders – and, after a final look to see that everything is right, we are ready to start. (The quickest start I have ever made was just two minutes and twenty-seven seconds, from the time I opened the first air valve until we were running, with the clutch in.)

Now, which way? Remember that the Kahlenberg runs either ways, unlike many oil engines, which are uni-directional. Ahead? All right!

Get the injection control lever in mid position, the throttle about half way open, and move the air control handle forward, holding it not more than a second, move it back to neutral, and, as smoothly as a steam engine, she has started. When new and stiff, about 100 pounds of air are required to start a four-cylinder engine, with a little less for the three and two. When well run in, any of the engines can be started with 60 pounds. Note that there is no missing or "cutting up" as you probably have been accustomed to when starting two stroke engines, or other oil engines, where the governor does not function properly. Each power impulse of each cylinder is produced by the injection of a measured amount of fuel.

Now, let us quickly pull in the clutch. Don't "nurse it in", jam it in quickly.

Suppose that the engine is running 350 R.P.M. There are four power impulses per revolution, or approximately one every quarter-second. So, when there is a demand for power, by the propeller, the first quarter-second later the governor has lengthened the stroke of the fuel pump for that cylinder next to function, and instantly has begun the delivery of the proper amount of fuel to that particular cylinder, and to the rest, in turn, as long as power is required. It seems almost incredible, but, before the flywheel has revolved once, since the load was thrown on, the entire four cylinders are getting exactly the right amount of fuel to hold the engine speed exactly at 350. In twenty years of engineering work, I have never seen a more sensitive governor on an engine of commercial type.

You can set the governor instantly at any speed between 90 and 400 and throw the clutch in and out all you please and still your ear can detect no material change in speed. Of course, the two-cylinder engine will not throttle quite as low as the three or four, but I have slowed a two cylinder down to 112 revolutions, and it functioned perfectly.

If now, we want to reverse, we have three methods.

First, we have the reverse gear built especially for the Kahlenberg engine. It is an extra heavy duty type, designed to carry much more than the rated capacity, and with a back-up speed 90% of the ahead.

Second, if we prefer reversing with air, we simply throw the throttle lever to the extreme forward position, entirely cutting off the fuel, and the engine quickly comes to a stop. Then half open the throttle, move the air lever aft, just an instant, then return it to neutral, and we are ready for a long pull astern. Remember that the Kahlenberg engine runs right hand exactly the same as left, so we can back at full speed indefinitely with ill effects. Think what that means if anything should happen to the reverse gear.

A third way of reversing is to throw the throttle hard ahead, entirely cutting off the fuel. The engine immediately starts to slow down. Set the injection control for full lead, in the direction of rotation, then when the engine is down to about 60 revolutions, quickly open the throttle wide. This will cause a full injection of fuel on the up-stroke, and a pre-ignition, which will reverse the engine positively and quickly. This method is used on several Continental engines.

Now that the engine is running, look her over. Note the smooth action of the governor and fuel pumps. The governor driving gear runs in an oil bath; a little oil added every week or ten days will keep the level right. The fuel pumps are up where they are accessible; furthermore, should trouble develop in any pump or fuel line, the pump is locked out of action by pressing down on the hand plunger over the pump and twisting to the right. Then by taking out two cap screws, the pump may be removed and a spare pump substituted, should it be found inoperative. The pump actuating mechanism also runs in an oil bath; thus it is impossible for it to rust up and stick or "freeze". The governor weights are also enclosed. In fact, the only moving parts visible on the engine are: the governor, weight casing, two short lengths of governor shaft, the flywheel, the eccentrics and connecting rods of the two water pumps and the air compressor, and the connecting links to the lubricators. There are two air intake valves to each cylinder base; these are wonderfully quiet, as the result of the use of tempered bronze instead of rubber or leather commonly used.

Note the system of handling the fuel. Oil flows to the pumps from a service tank, which is fitted with a double strainer having two and one-half square feet of surface. Just before reaching the pumps, the fuel passes through a smaller and finer strainer, and in each individual fuel line, there is a final strainer, just below the nozzle. This last is an exceedingly ingenious device, passing the fuel through holes smaller than the injection nozzle orifices, making it impossible for dirt to clog them. This attention to detail indicates the care with which this engine has been worked out to the end that every possible chance for trouble has been eliminated.

Look now, at the after end of the engine. There is a circulating pump on one side and bilge pump on the other, identical and interchangeable. If a part of the circulating pump should happen to break, you can replace it with a part from the bilge pump. Or, if desired, you can cross-connect, so that the bilge pump can be used in emergency as a circulating pump. A large air chamber prevents hammering.

The air compressor is mounted on the rear cylinder above the shaft, on the mid-line, and is water-cooled.

There are two force-feed oilers. The port oiler takes care of all the crankshaft bearings and the thrust bearing. In this oiler comparatively cheap oil is used, as it is not necessary to use an oil with the high fire test required for cylinder lubrication. The starboard oiler takes care of the cylinders – each cylinder is oiled from two sides, and the wrist pin through its end, by a special line. The crank pin is likewise oiled by force-feed through an oil ring fixed to the crank web, the crank pin being hollow. By installing lubricating oil tanks of sufficient capacity above the oilers, and piping them direct to the lubricators, it is possible to provide almost automatic lubrication for several hundred hours of running.

There is nothing more to the engine. It's just as simple as that. There's nothing on the engine to adjust. It's a beautifully finished machine – there is no finer workmanship on any European or American engine, inside or out, and few, indeed, are in the same class.

Is this oil engine dirty and "smelly"? Is it suitable for installation in a yacht? With the fuel lines properly set up the engine room can be kept absolutely sweet and clean, and while fuel oil is not desirable as perfume, there is no excuse for having any odor in the boat. Unlike the gasoline engine, where the fuel is vaporized and mixed with air, there are no vapors developed consequently no odors.

An engineer should keep the bright parts of the engine oiled, and the brass polished. There is no excuse for not doing so, even in a working boat. I have seen engines of various makes in fishing and other commercial boats, and the brass was shining like a mirror. I have seen other engines a mass of rust after six month's use. An oil engine is not a cheap installation. A Kahlenberg is as perfect a piece of mechanism as can be purchased. Its life, and the satisfaction derived from it depend, in no small way, on the care given it, and it is an economic crime to install one and then neglect it. If kept clean and thoroughly lubricated, it should run year after year without up-keep expense.

Does the engine run hot? A frequent question. After ten hours at full load, you can keep you hand on the cylinder walls or head without discomfort. Notice the huge exhaust manifold. This is really a big receiver where the products of combustion, immediately where the exhaust ports open, can expand, removing all possible back pressure from the piston. Thoroughly water jacketed, the temperature of the manifold walls should not exceed 120 degrees.

Is the exhaust noisy and does the engine smoke? Properly installed, the exhaust is almost noiseless, and, properly handled, there is absolutely no smoke. If there is smoke, it is either due to excessive lubrication, or failure of the engine to completely burn the fuel, and that is one thing the Kahlenberg lays claim to – complete combustion. This must be a fact, for only one engine of the same type, that I have been able to discover, is covered by as low fuel guarantee. This one hundred horse runs in every day service, hooked up to full load, on exactly seven gallons an hour, which is, with the fuel I am using .52½ pounds per horse power hour. The factory guarantee is .55 pounds, which could easily be made, running under constant load conditions. Then too, I strongly suspect that at my "hooked up" speed, the engine develops better than 110 horse power.

Now, just a little about the materials entering into this engine:

The crankshaft is a high carbon steel, double heat treated. Much of the reference to heat treatment in catalogs is "bunk", but every Kahlenberg crankshaft is double heat treated to produce the required tensile strength and other physical characteristics necessary in a good internal combustion engine crankshaft. The forgings are from the largest and best equipped steel mills, produced to Kahlenberg specifications.

- All crankshaft and crank pin bearings are of the highest quality white metal sweated into heavy bronze shells. A crank pin bearing for the 50 horse power two cylinder engine weighs thirty-four pounds.

- Cylinder castings are of the finest close-grained special grey iron and pistons are made from the same material.

- Connecting rods are forged steel and finished all over. Crank pin bearings are of the familiar marine type. Wrist pins are hardened tool steel, fitted to the piston, with bronze bushing in the connecting rod.

• All of the fittings are up to the standard of the engine itself. Force feed oilers, gauges, and similar parts are purchased from the best makers. Practically all the detail fittings are Kahlenberg design and make.

• Taken as a whole, the engine is a magnificent example of careful design and painstaking construction, and if you want a real sample of Kahlenberg enthusiasm, just get in touch with a Kahlenberg owner or operator – you'll get enthusiasm, to the nth degree.

The foregoing is a little yarn about the Kahlenberg heavy oil engine by an owner and operator of same, and I told it in everyday language, giving you the information naturally desired by a prospective buyer.

Appendix C

The Hicks

The following is reproduced from a data sheet supplied by Hicks.

Propeller

A Hicks Manganese Bronze propeller, "the result of a quarter of century practical experience with commercial fishing boats and work boats," so designed to give the maximum speed.

Water Circulating System

All fittings necessary for the intake and discharge of cooling water

Battery

6-Volt – 100 Amp. Hr. with moisture proof spark coil for auxiliary ignition.

Air Tank and Whistle

Galvanized steel tank and Organ Whistle with complete installation equipment from the air compressor to whistle.

Fuel Tanks

Extra heavy galvanized steel, including all fittings from the deck plate to the carburetor.

Installation Equipment

Complete engine installation equipment provided.

Bilge Pump

A specially designed positive acting rotary pump which delivers water at high pressure, suitable for washing decks, fighting fire, etc. It has a 3-way cock so as to pump out bilge or supply water from outside for washing purposes.

Generator

Attachment whereby the standard automobile type of generator can be driven from the flywheel.

The following advice was provided by Hicks to those who purchased its engines:

Important!

When Installing a Hicks Engine Be Sure That:

1. All gasoline tank and pipe connections are tight.
2. The vent hole in the filler connection to the gas tank is drilled.
3. All nuts on the installation lag studs are tight.
4. There are no air leaks in the intake manifold.
5. The bleeder pipe on the exhaust manifold is open.
6. The drain holes in the bedplate are open.
7. There is sufficient oil in the crankcase for splash lubrication.
8. The force feed oiler is full of fresh oil.
9. The air compressor sight feed oiler is full of fresh oil.
10. All oil feeds are open sufficiently to insure ample lubrication.
11. The oil wells on the thrust bearings and flywheel bearings are full of fresh oil.
12. The camshaft bearings are well oiled.
13. The oil wells in the reverse gear pins are full of fresh oil.
14. The water pump and air compressor eccentric straps are well oiled.
15. The water pump is functioning.
16. You can turn the propeller and propeller shaft with your hand before it is coupled to engine with the intermediate shift.

To Start a Hicks Heavy Duty Marine Gas Engine:

1. Retard spark completely by pushing spark lever forward.
2. Open throttle half way. Throttle handle is located on aft end of rocker arm assembly.
3. Open exhaust relief just sufficiently to permit the engine to be turned over. Do not open the exhaust relief all the way. Exhaust relief handle is on forward end of rocker arm assembly.
4. Prime all cylinders with gasoline through priming cocks on cylinder heads.
5. Depress the float on the carburetor with tickler to make sure that there is sufficient gasoline in the carburetor. This operation should flood the carburetor.
6. Turn the flywheel until a spark rod tripper comes in contact with an igniter pawl on any cylinder. Then give the flywheel a quick turn past the point where the igniter fires.
7. As soon as the engine starts, close exhaust relief and advance the spark. If the engine does not pick up rapidly, hold the hand over the air intake horn on the carburetor. The engine speed is regulated with the throttle handle.
8. Permit the engine to obtain a moderate speed before throwing in the clutch.

Index

About the Author

Stan Grayson's work in the field of nautical and automotive history is well-known to readers. Widely recognized for his ability to bring alive the technology of an earlier era and combine it with the stories of the men and companies who created it, Stan spent over four years researching and writing *Engines Afloat, from Early Days to D-Day.* He is the author of six books, has edited numerous others, and has written articles for a variety of magazines including *Automobile Quarterly* and *Nautical Quarterly.*